Laser Cladding for Restoring and Increasing the Durability of Railway Wheels

Zaure Zhetpisbaevna Zhumekenova
Vitaliy Vladimirovich Savinkin
Andrei Victor Sandu
Petrica Vizureanu

Reviewers:

Doctor of Technical Sciences Kuznetsova V.N., Dean of the Faculty of Oil and Gas and Construction Equipment (Siberian State Automobile and Road Engineering University, Omsk, Russia);

Candidate of Technical Sciences S.B. Musrepov, Director (North Kazakhstan College);

PhD A.A. Kashevkin, Head of Chair, Associate Professor of department of energetic and radioelectronics(Non-profit limited company «Manash Kozybayev North Kazakhstan university»)

Published as part of the book series
Materials Research Foundations
Volume 157 (2024)
ISSN 2471-8890 (Print)
ISSN 2471-8904 (Online)

Print ISBN 978-1-64490-290-5
ePDF ISBN 978-1-64490-291-2

Distributed worldwide by

Materials Research Forum LLC
105 Springdale Lane
Millersville, PA 17551
USA
https://www.mrforum.com

Printed in the United States of America
10 9 8 7 6 5 4 3 2 1

Table of Contents

Preface

The monograph examines and explores the issues of increasing the durability of wheel sets of railway cars, the main causes of failures and wear of wagon wheels due to intensive wear of critical components, parts. The results of the strength calculation of the main loaded elements of the wagon wheel using the Solidworks software using the Static II Pro package for the analysis, visualization and study of fatigue stresses on the contact surfaces of the wheel and rail are presented. To ensure the durability of the rolling surface of railway car wheelsets, it is proposed to use laser surfacing technology with the development of a mobile repair complex for restoring wheels with a laser energy source that provides adaptive control of thermodynamic processes of surface and structure formation.

The monograph is intended for bachelors, masters, doctoral students and specialists in the field of mechanical engineering and energy, and can also be used by employees of scientific, design, production and repair and restoration enterprises.

Designations and Abbreviations

MP – maintenance point

WPS – wheel pair set

RS – rolling stock

MMMM – metal magnetic memory method

NDT – non-destructive testing

SCZ – stress concentration zone

RMF – residual magnetic field

ECM – eddy current method

HAZ – heat affected zone

LIZ – laser impact zone

NPC – numerical program control

Introduction

Due to the geographical features of the Republic of Kazakhstan, within the continent, in the center of Eurasia, railway rolling stock, track machines and special-purpose complexes are actively used to ensure transport communications and cargo turnover. In the Republic of Kazakhstan, railway capacities form the infrastructure and ensure import and export of the national economic market. The railway fleet in Kazakhstan for 2021 averages 1.7 thousand locomotives, electric and diesel locomotives, 291 automated machines and complexes, 2.7 thousand passenger cars, 28 thousand baggage cars, 54.9 thousand freight cars and 75.5 thousand cars owned by private companies. Given this volume, about 65% of the railway fleet is morally and physically worn out, which reduces their efficiency and economic feasibility of use. The Tulpar-Talgo LLP plant existing in Astana has a narrow specialization in the production of passenger cars; its capacity does not provide Kazakhstan with a full turnover of the railway freight fleet, which indicates the relevance of the topic of ensuring the durability of railway cars using technological methods.

Currently, large industrial enterprises have railway sidings in their facilities, as well as railway repair units. Operation in different climatic zones and the functional features of rolling stock lead to intense wear of critical components and parts and, as a consequence, to a sharp decrease in the service life and safety of the railway stock. Dynamically active systems, especially the chassis of railway cars, are subject to the most progressive wear and loss of reliability. Due to the high cost of new wheel pairs from 1,5 million tenge to 3,5 million tenge, and the unstable level of inflation, it is proposed to restore worn car wheels as an alternative to new ones. There is a shortage of enterprises that can provide restoration of carriage wheels in Kazakhstan, and they carry out repairs directly on the basis of their own organization. The issue of effective restoration of wheel sets at a distance from repair points along the route has not been fully studied. Existing restoration methods do not meet the requirements for the quality of the modified surface, and the physical and mechanical properties are lower than those of new wheelsets. Thus, the scientific and practical problem of eliminating defects and carrying out major repair and restoration work at a distance from depots and repair bases has not been solved. The scientific problem is the lack of substantiated methods for ensuring the durability of railway wheels in the field using highly concentrated sources of laser energy and substantiated optimal technological recovery modes.

A method has been developed for calculating the strength characteristics of railway car wheels, taking into account the deviation of the contact patch of the worn part of the surface under the cyclic distribution of impact dynamic loads and axial moments. The introduced technique makes it possible to determine the zone of localization of fatigue stresses and predict a defect before its detection. This approach increases the manufacturability of the structure during its design, the reliability of the dynamic system and increases the overhaul life of the wheelsets.

The developed methodology and algorithm for ensuring the durability of railway wheels makes it possible to substantiate the efficiency criteria of wheel pairs of railway cars. Well-founded criteria of the proposed algorithm provide a reliable choice of restoration method, and the cause-and-effect relationships of defects established in it create the possibility of optimal selection of technological modes for restoration or modification of a worn wheel tread.

A virtual simulation model has been proposed in the Solidworks environment to study the process of stress localization along the contact elements of the wheel at different intervals of exposure to shock loads, which allows one to estimate and clarify with satisfactory convergence the magnitude and principle of distribution of contact stresses along the complex profile and flange of the wheel.

The developed mobile repair complex for the restoration of wheel pairs of railway cars allows you to quickly restore worn wheels in the field at a distance from the repair depot. Its unique repair booth platform provides a process environment and conditions similar to those found in fixed repair stations. The manufacturability of the proposed equipment allows you to automate the process and reduce its energy intensity.

The introduction of the developed new mobile complex into the infrastructure will solve the technical problem of increasing the overhaul life of wheels. The developed mobile complex with a laser energy source ensures restoration with any complex and refractory material. Laser surfacing provides the width of the surfacing weld depending on the diameter of the nozzle and allows flexible control of the physical and mechanical properties of the coating. The power of the laser pulse provides a targeted effect, eliminating overheating of the entire part, which leads to scorching. The use of a laser energy source in the mobile complex will ensure effective control of thermodynamic processes during the formation of the surface and structure of the wheel. The studied optimal technological parameters of restoration improve the quality of the coating, eliminating tensile and compressive stresses.

Established dependences of changes in the physical and mechanical properties of the restored surface on the technological modes of laser restoration. These dependencies make it possible to reasonably select the optimal values for the restoration modes of the worn tread surface and wheel flange. The variability of reasonable restoration parameters affects the formation of optimal microhardness of the coating, high adhesion strength and increases wear resistance, which leads to an increase in the service life of car wheels.

The developed algorithm and technological method for laser restoration of a worn wheel increases the variability of technological restoration modes, expands the horizon of possibilities for improving and modifying the physical and mechanical properties of the modified worn wheel surface. The uniqueness of the proposed laser source is adapted to the harsh conditions of restoration at a distance from repair bases and makes it possible to achieve a thermally stable surfacing process with a minimum depth of penetration of the base metal in a short period of time. The developed restoration method increases the list of products subject to restoration.

Scientific and practical problems have been solved aimed at increasing the reliability, durability and efficiency of shaping the tread surface of rolling stock wheel pairs and allowing us to fundamentally change the concept of repair and restoration work of a railway stock without taking the car out of service along the train's route.

Chapter 1. Analysis of Design Features of Wheel Pairs of Railway Cars and Prospects for Increasing Their Durability

1.1 World practice of developing and improving the design of wheel pairs

Carriage wheel pairs provide direct contact with the part of the railway unit, traction, chassis, track and guide the rolling stock in the rail track. The reliability, stability and safety of wagon movement depend on the design accuracy of geometric dimensions, physical and mechanical properties, strength, and pairing of wheel pairs. The rolling stock wheel pair is shown in Figure 1, in the form of an axle and two steel wheels pressed onto it [1].

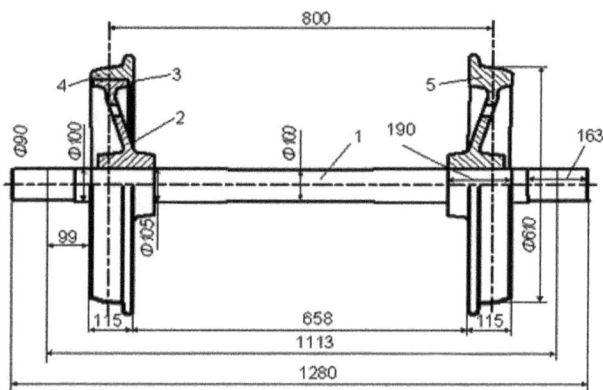

1 - axis; 2 - disk center; 3 - safety ring;
4 - bandage; 5 - solid rolled whee

Figure 1. Bandage wheel pair of rolling stock cars

The intensive development of the power of the railway fleet is formed by increasing the mass and speed of rolling stock, which leads to an increase in power loads from the car body to the bogie with a wheel pair, then to the rails. With an intensive increase in workloads from the car wheel on the rail track and an increase in travel speeds, the resistance forces during the contact interaction "wheel-rail" increase and the fatigue-stressed state of the metal of the rails and wheels increases. In recent years, the railway track in Kazakhstan has undergone significant modernization [2]. The operation of rolling stock in areas with low temperatures, especially in the north of Kazakhstan, has complicated the work of wheel sets. Therefore, research is constantly being carried out in Kazakhstan aimed at improving the design and quality of the material of carriage wheels.

Due to the peculiarities of design solutions, carriage wheels are divided into one-piece (one-piece); hub (bandage - consisting of a wheel center, a bandage and a safety ring); elastic (containing an elastic element between the tire and the wheel center). Based on the manufacturing method, carriage wheels are divided into rolled and cast. According to the dimensional geometric features of the wheel diameter - 950 mm and 1050 mm, measured along the rolling surface [3].

Carriage wheels are operated in various harsh climatic conditions and take on significant static and dynamic loads through a small contact area. At the same time, when braking, friction forces arise between the wheels, pads and the rail track, causing heating and wear of the wheel, and as a result, the appearance of defects. Movement on curved turning areas gives rise to chips, cracks, and spalls in the wheel rim. This phenomenon affects the safety of railway traffic.

When studying the design of wagon wheels, the following factors were derived that ensure durability: the metal of the treadmill should have high hardness, wear resistance and strength, the metal of the disc and wheel hub should be moderately viscous. The obtained conditions support non–target wheels, where the bandage is made of steel of increased hardness and strength, the wheel center is made of relatively viscous and cheap steel [4].

A standard solid-rolled steel wheel (Figure2, a) is made of the working part of the wheel - a rim with a rolling surface, a disk and a hub. The wheel has an inner edge and an outer edge. The test rolling surface is ground according to a standard profile. According to GOST 10791-2011, solid wheels are made from steel grades 1 and 2 [5]. The differences in the chemical composition of steel grade 1 from grade 2 is in the carbon content; the percentage of carbon of grade 2 is higher than that of wheel steel 1. The mechanical properties of wheel steel after heat treatment must correspond to the nominal values (Table 1).

Table 1. Optimal parameters of wheel steel after restoration

Metal brand of the wheelset	Relativeelongation, noless	Relativenarrowing, noless	Surfacehardness, *HB*	Tensilestrength, MPa	Brittleness at 20°C, MJ/m^2
Steel 1	12 %	21 %	248	880-1080	0,3
Steel 2	8 %	14 %	255	912-1107	0,2

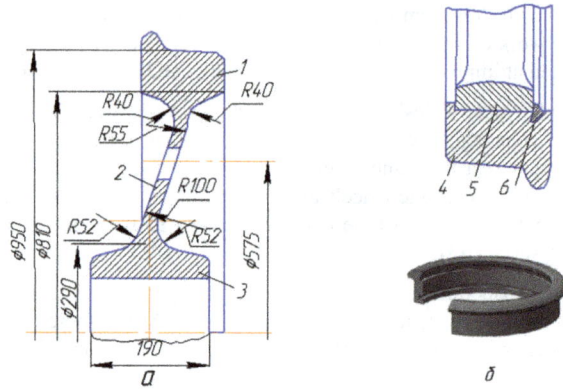

a — bandless solid-rolled; b – bandage
1 - rim; 2 – disk; 3 – hub; 4 - bandage; 5 – wheel center; 6 – reinforcing ring

Figure 2. Profile of a standard car wheel

The design of the tire wheel (Figure 2, b) consists of a removable tire, a wheel center and a safety ring. Bandage wheels have a dimensional diametrical differencebetween the tread surface and the hub.This design type of wheels is widely used in the market of near and far abroad.

The design of tireless solid wheels is much stronger and more reliable than tires in such parameters as (loosening of the tire, frequent formation of wheel failures in the form of cracks and wheel displacement from the axle), production costs for the formation of a wheelset (boring and fitting of tires), weight (wheel with a diameter of 950 mm is 36 kg heavier than the diameter of 1050 mm). The presented parameters are mainly noticeable when the train speed increases and the load on the wheel center increases. As a result, tire wheels are in some cases replaced with the optimal option for solving a technical design problem as tireless ones [6-7].

The main design feature of wheels is the different profiles of the wheel tread surface. When contacting the "wheel-track" system, the profile of the car wheel has a noticeable effect (Figure 3). The ridge guides the wheel along the rails, and an increase in the angle of inclination of the ridge contributes to the stable movement of the wheelset on the rails and the least wear [8]. The nominal standard parameters of the 1:10 taper of the tread surface profile on a straight section of the track prevents uneven wear of the wheel rim width and helps when passing at curved track joints, but creates a tortuous movement, negatively affecting the smooth running of the rolling stock [9-12].

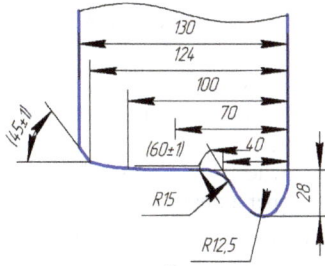

Figure 3. Wheel tread profile

The design of elastic carriage wheels is divided into carriage wheels with rubber gaskets (Figure4) and with pneumatic tires. The design features of elastic wheels are aimed at reducing the impact of impact forces and high-frequency vibrations, improving the smoothness of the ride and reducing the noise that occurs during its movement [13-15].

Thus, modern production of wheel sets, to ensure high reliability, imposes strict requirements on operational parameters [16].

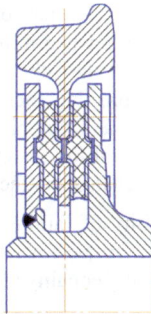

Figure 4. Design of an elastic carriage wheel with a rubber gasket

From analytical studies it has been established that the most used railway wheels are solid-rolled tireless wheels. It is more economically feasible to restore tireless wheels than to discard them, therefore the development of a method and technological equipment for restoring the tread surface of a wheelset is a priority task.

It has been established that it is necessary to consider the wheel not as the only element of the railway system, but as a complex integral structure of all interconnected units with nominal geometric parameters. The scientific task is to establish the optimal dimensional parameters of the wheel.

When determining the geometric parameters of the wheel, it is necessary to take into account the vector of loads, the dynamics of the movement process, the properties of the metal and the quality of the machined surfaces. The proposed approach to systematizing research and justifying optimal wheel parameters will help solve the complex problem of increasing the efficiency of the operational characteristics of structural elements [17-19].

The basic requirements that increase the reliability of wheel operation have been identified. The main design and geometric parameters of the wheel have been established, ensuring high operational efficiency of the wheels: increasing the angle of inclination of the ridge contributes to the stable movement of the wheelset on the rails, the least wear, reducing damage to the rolling stock and the severity of the consequences from them.

The performance qualities of wheels are characterized by their ability to withstand cyclic stresses resulting from an increase in the speed of rolling stock, wheel loads and severe braking conditions, reaching the limit of safe operation.

1.2 Methodological approaches in assessing the reliability of wheel sets of railway cars, diagnosing defects and studying the wear process

Ensuring the durability, efficiency, strength and quality of the wheelset is ensured by their reliability and trouble-free operation in difficult operating conditions. Reliability depends on the complex performance parameters of the wheel pair ($P(T_p)$, $ñ$, N_{sum}, F, $F(t)$, $P(\tau)$) and its intensely stressed state when moving under the influence of constant mechanodynamic loads, leading to damage and destruction. Therefore, determining effective diagnostic methods and establishing their cause-and-effect relationship is an urgent task in predicting the reliability and durability of a wheel. The accuracy of the prediction will depend on the time interval for detecting the malfunction and the method of diagnosing the defect and structural changes in the car wheel.

Based on $P(T_r) = 0{,}99$ - the probability of failure-free operation of the car wheel, we determine the absence of unacceptable fatigue cracks in the wheel elements in accordance with GOST 33783–2016 by the following equality (1):

$$P\left(T_p\right) = 0{,}5 + F\left(\frac{{}^{n_p}/_{n}-1}{\sqrt{({}^{n_p}/_n)^2 \cdot v_{\sigma-1D}^2 + v_{\sigma a}^2}}\right). \tag{1}$$

where ${}^{n_p}/_{n}$ - the relative safety factor [20 - 22];

$v_{\sigma-1D}=0{,}06\text{-}0{,}08$ – coefficient of variation of the endurance limit (GOST 33783-2016);

$v_{\sigma a} = 0{,}1\text{-}0{,}15$ - the coefficient of variation of dynamic stresses (GOST 33783-2016);

$F(t)$ - the Laplace time function or the probability integral [23].

We select the maximum load factor depending on the degree of fatigue curve $m=6$ according to GOST 10791-2011 [24-25]:

$$n_p = 2{,}797 - 0{,}9294 \cdot \left(\kappa \cdot \frac{N_{sum}}{N_0} \right)^{0{,}05}, \tag{2}$$

where N_0 - the test base (GOST 33783–2016);

   $k-$ intensity factor for reducing the wheel endurance limit, $k=1{,}65$ (GOST 33783–2016).

We determine the total number of wheel loading cycles for the designated service life T_r, by equality:

$$N_{sum} = \frac{365 \cdot T_p \cdot L_i}{P \cdot D}, \tag{3}$$

where L_i - the average mileage of rolling stock per day;

   D - the average diameter of the worn wheel.

We determine the probability integral depending on the time interval using equality (4):

$$F(t) = \frac{1}{\sqrt{2\pi}} \int_0^t e^{-\frac{t^2}{2}} dt. \tag{4}$$

The maximum load factor based on average values is found from the equality:

$$n = \frac{1 + 0{,}1 \cdot U_{pmax}}{n_{y.\kappa}}, \tag{5}$$

where $U_{p\,max}= 5\text{-}5{,}5$ – quantile of normal distribution (GOST 33783–2016); $n_{y.\kappa}=n_{min}=1{,}3$ – fatigue resistance safety factor (GOST 33783–2016).

A graphical representation of the probability of a random variable falling into the intervals $[0, t_1]$, $[t_1, t_2]$ and $[t_1, \infty]$ is shown in Figure 5.

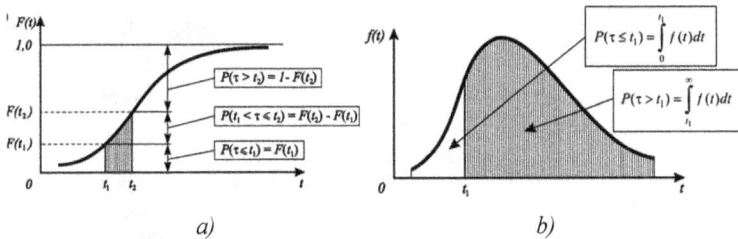

a – integral function F(t); b – differential function f(t)

Figure 5. Study of the probability of a random variable falling into a given interval

The differential distribution density function $f(t)$ describes the distribution law of a continuous random variable. The differential function characterizes the density with which the values of a random variable are distributed at a given point [26].

Function $f(t)$ is the derivative of function $F(t)$, so $f(t) = F'(t) = dF(t)/dt$. This means that to determine the probability of failure $F(t)$, we integrate the probability density function $f(t)$:

$$F(t) = P\{\tau \le t\} = \int_0^t f(t)dt. \tag{6}$$

The differential distribution curve of time to failure is depicted by the function $f(t)$.

Next, we determine the probability of failure during operating time t_1 (the probability that the random variable will take a value $\le t_1$):

$$P\{\tau \le t_1\} = \int_0^{t_1} f(t)dt, \tag{7}$$

- probability of failure-free operation for operating time t_1 (probability that the random variable will take the value $> t_1$):

$$P\{\tau > t_1\} = \int_{t_1}^{\infty} f(t)dt, \tag{8}$$

The variance has the dimension of the square of a random variable, which is not always convenient.

Standard deviation of a random variable

$$s = \sqrt{D[\tau]}. \tag{9}$$

The coefficient of variation is a relative measure of the dispersion of a random variable

$$v = \frac{s}{m_t}. \tag{10}$$

Therefore, to assess the real level of reliability of car wheels, it is advisable to use quantitative characteristics - reliability indicators.

Based on the results of the analysis, it was established that in existing methods for predicting failure, the time interval of observations and the number of defects for the period under study is taken as a basis. The action of forces and moments on the wheels of the car is interpreted in a general form and is represented by the number of loading cycles N_{sum}, related to the technical characteristics of the steel grade of the wheel [27]. This criterion is of a formal nature for assessing the reliability and durability of a car wheel and does not in any way characterize the impact of a cyclically changing dynamic load, which is a significant drawback in the forecasting theory. Also, classical methods for predicting complex technical systems, for example, «wheel pair-car», do not take into

Laser cladding for restoring and increasing the durability of railway wheels   Materials Research Forum LLC
Materials Research Foundations **157** (2024)   https://doi.org/10.21741/9781644902912

account the structural arrangement of elements, their axial displacement and natural progressive wear during operation.

Thus, a scientific problem has been formulated about the need to improve the methodology for accurately predicting the reliability of complex dynamic systems, for example, «body - bogie - wheel – rail», taking into account progressive wear and changing dynamic characteristics. This problem is proposed to be solved by introducing into the mathematical reliability model a criterion that takes into account the change and deviation of the design axis of the contact defect of the wear surface.

The next factor affecting the service life of loaded carriage wheels and the diagnosis of defects is the wear of resource connections and matings; accumulation of fatigue damage in car wheels; change in the physical and mechanical properties of the wheel metal. The main sources of defect generation in energy-loaded structures are stress concentration zones (SCZs). SCZ are zones of dynamic and mechanical action in which the processes of fatigue, corrosion and creep develop most intensively. The danger of SCZ is that fatigue processes occur very slowly, hidden not only for visual examination, but also for many diagnostic devices. The consequence of failure to detect microcracks and structural and phase changes in the material of the loaded part are failures and accidents [28]. Elimination of the consequences and restoration of the original manufacturability and constructibility of carriage wheels requires additional unplanned financial costs.

In non-destructive testing (NDT) using forced magnetization, a special place is occupied by the metal magnetic memory (MMM) method. The MMM method records the distribution of the intensity of the self-residual magnetic field (RMF) of steel parts[29,30]. Researchers of the MMM technique determine that identifying the location of the SC zone is an important solution to finding the source of intense physical-mechanical and chemical-thermal defects of the metal. At the same time, the nature of the change in magnetization and the level of tension of the magnetic field is not dependent on the physical and mechanical structure of the metal, but reflects the actual stress-strain state, which is probabilistic in nature. The device used in the MMM method for detecting faults is a flaw detector of the IKN-3M-12 brand.

Significant factors influencing the nature of the WMD are the structural class of steel, depending on the chemical composition (low-medium carbon, low-medium alloyed, high alloyed), the type of initial state (hot rolled, normalized, thermally improved) and the type of technological processing (rolling, welding, casting, etc.).

The main disadvantages of the method include low resistance to external influences, but the main problem is that it is not known which zone of scattering of the OMF tension corresponds to the zone of concentration of mechanical damage from fatigue defective changes in the wheel metal.

One of the common NDT methods in the study of deep defects in carriage wheels is ultrasonic (ST RK 1675-2007). The main disadvantage of the ultrasonic method is the detection of defects at a depth of $h \geq 1$ mm, the relative error h is $\approx 25\%$, which negatively affects the assessment of the degree of danger of a defect.

When exploring the capabilities of the eddy current method (ECM), it is necessary to evaluate the adequacy, accuracy and sensitivity of magnetic fields to different grades of material, depth and nature of the defect. For non-destructive testing of carriage wheels, it is possible to use a VD-113 eddy current flaw detector.

It has been established that the main task of non-destructive testing is not only to assess the location and cause of defects, but also to obtain accurate measurement results with the smallest error. This problem can be solved using a multi-vector approach, taking into account system connections in different planes of the problem [31-33].

Scientific and technological needs are concentrated in the absence of domestic technologies beyond accurate forecasting of critical parts of mechanical engineering in Kazakhstan.

Based on the results of diagnostic studies with the recommended cross-method of assessment (MMM and VTM) of internal stresses and structural changes, it was possible to expand the range of defects that previous researchers had not previously taken into account when defecting highly loaded parts, but made a significant contribution to the development of dynamic loads, reducing service life and operational safety car wheel pairs. The listed features expand quality control in the production and technological processes of repair and restoration production of wheel sets[34-36].

Based on the results of studies of modern non-destructive testing methods, diagnostic systems and technologies for restoring the geometry and properties of highly loaded parts, the most effective methods were reasonably selected and recommended. Clear criteria have been formulated that characterize the conditions for possible adaptation of diagnostic methods and recovery methods to real production conditions. Factors have been identified that influence the full integration of non-destructive testing methods, regulating the quality and cost indicators of work[37-40].

From the theoretical studies carried out, it was revealed that the MMM method, the eddy current method and the integration of predictive calculations are promising as an element of a set of methods for the comprehensive assessment of rolling stock wheelsets. A comparative analysis of errors in measuring defects of all methods of non-destructive testing of car wheels showed the following values: radiographic - 13 – 30%, VTM -10-30 %, ultrasonic - 3-15 %, MMM – 0,5-3%.

Using methodological approaches in the field of wagon wheel wear, it is necessary to assess the wear of the wheel profile during the operation of rolling stock. There are regression and probabilistic methods for statistical analysis of the study of car wheel wear [41].

Regression analysis consists of periodic quantitative statistics of surface wear of the studied wheels of railway cars. We select 3-5 cars of the same train and on the same route, while the wear under study is measured every 10-20 thousand kilometers. Then, after 100-200 thousand kilometers, the measurement results were generalized to make conclusions.

According to the results of the study, tread surface defects were found in several wheels of the rolling stock after 170 kilometers. Among other studies, regression studies of tread wear over a mileage interval from 130 to 170 km were completed. The results of a regression study of wear on the rolling surface of a rolling stock wheel are presented in Figure 6.

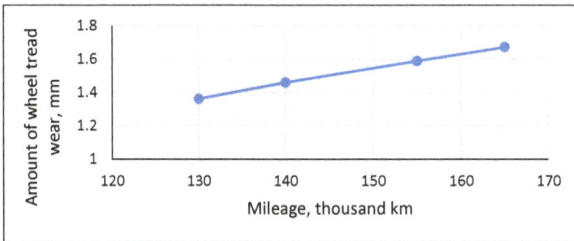

Figure 6. Results of studying the wear of the rolling surface of a rolling stock wheel using the regression method

Analyzing Figure 6 it can be seen that the average wear of the running track of a carriage wheel increases with increasing mileage, that is, from 1.36 to 1.67 mm for a run of more than 40 thousand km, averaging 0.13 mm per 10 thousand km.

When studying wear using a probabilistic method, it is not necessary to monitor the complete wear process; only statistical data for a short period of time is required [42]. The results of a probabilistic study of tread wear are presented in Figure 7.

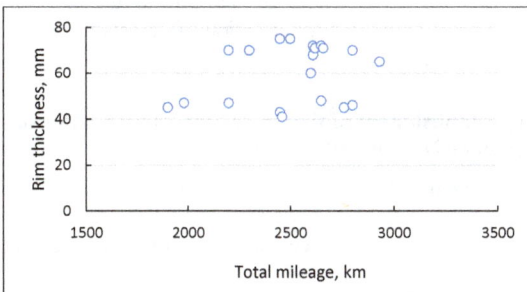

Figure 7. Results of studying the wear of the rolling surface of a railway car wheel using the probabilistic method

The change in the difference in wheel diameter between wheels of the same axle and flange wear depending on the total mileage of the locomotive is shown in Figure 8.

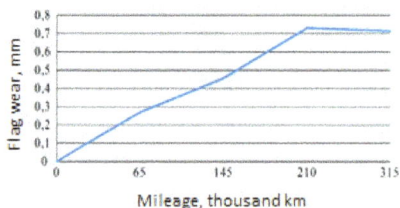

Figure 8. Change in flange wear

The advantage of the regression method is that it is systematic; it is possible to correctly diagnose critical moments during the operation of a carriage wheel: the transition point from repair-free mileage to trouble-free operation, then to increased wear. The advantageous capabilities of the probabilistic method lie in the insignificant amount of work, one-time, quick measurements, that is, the research period is short.

A limitation of the probabilistic method is the inconsistent detection of wear on each wheel. The disadvantage of the regression method is that a small number of wheels are studied using this method, although the content and period of the research work are large [43].

Consequently, both considered methods equally determine the estimate of the average wear of a rolling stock wheel profile, however, the best correlation between mileage and wheel wear is obtained using the regression method and residual magnetization.

The defects studied by these NDT methods are different in type, shape and location; in particular, damage directed at an angle of ≈40-50° to the rim surface is easier to determine. The physical meaning of the formation of defects has not been fully studied, so the next task is to establish the cause-and-effect relationship of wheel defects.

1.3   Research and establishment of the cause-and-effect relationship of wheel pair defects and assessment of the severity of their consequences

When studying the wear of wheels in operation, more than 60 damages are observed. Research of scientific works of M.M. Mashneva, A.E. Tsikunova, A.F. Bogdanova et al. are aimed at separating defects in wheel pairs according to the elements of carriage wheels [44-46].

All malfunctions of wheel sets and components are classified as follows: malfunctions of wheels, axles and wheel sets, in particular the treadmill. The studied damages can be classified into: metal continuity defects, wear defects, fatigue defects, fracture defects and damage[47].

The amount of wheel wear depends on the condition of the track, contact conditions during friction, physical and mechanical properties of the metal and other factors. In many ways, the efficiency of car operation will depend on the structure of the defect and the limits of its possible values that do not reduce operational performance (V_v, t_{put}, t_{tor}, $S_{tor}$and etc.).Thus, a period of time inevitably comes when it is necessary to take the car out of service and place it for maintenance or repair. Qualitative reliability characteristics are regulated by GOST 32192-2013, exceeding which requires replacement with new ones [48]. As an alternative to replacing railway wagon wheels with new ones, we offer restoration using surfacing technologies. The repair process is very labor-intensive, costly and requires sound technological regimes, effective technologies and the application of scientific principles. It is technology that influences the formation of the physical and mechanical properties of the surface and affects technological defects. Analyzing the defect detection reports of the repair depot of JSC National Company Kazakhstan TemirZholy, the main defects and the reasons for their occurrence are formulated (Table 2) [49].

Table 2. Defects in railway car wheels and the reasons for their occurrence

Type of defect	Illustration	Cause of occurrence	Severity of consequences
1	2	3	4
FATIGUE DEFECTS			
Cracks			Changing the trajectory of movement «wheel surface – rail». Increased noise. Shock loads.
Kinks		Metal fatigue under cyclic loading	Car derailment. Technological disaster. Death of passengers, loss of cargo. Destruction of therailbed.
Metalchipping (gouges, shells)			Increased noise, rail wear, wheel slippage, operating inefficiency.

15

DEFECTS ASSOCIATED WITH WEAR		
Crawlers		Characteristic impacts leading to the destruction of wheels, ridges and rails. Inwinterconditionstherisk of destructionincreases.
	Changes in the shape and size of parts as a result of friction, jamming of the wheelset	
Growths		
DESTRUCTION		
Breakaways		Wheelalignment
	Due to fatigue destruction of the surface layers of the metal, thermal cracks, hidden metal defects	
DAMAGE		
Localrimwidening		Damage to track superstructure elements.
	Plastic deformation of metal under the influence of normal cyclic forces	
METAL CONTINUITY VIOLATION		
Cracks in the hub, longitudinal cracks		Destruction of wheel pair elements.
	Defects of metallurgical and rolling origin	

To study defects in wheel pairs, we will analyze the distribution of types of defects among the structural elements of the chassis of a railway car.

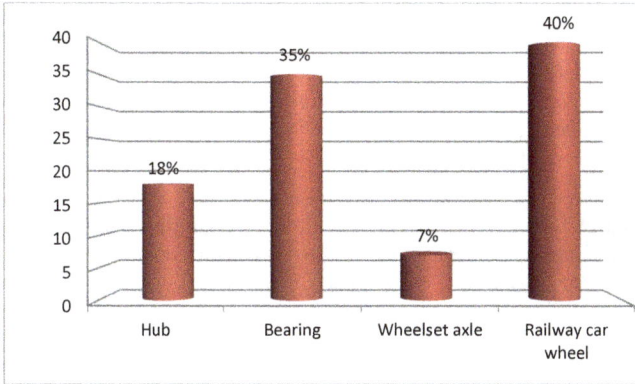

Figure 9. Failure modes for wheel pair designs

An analysis of Figure 9 shows that about a third of failures occur in bearings and car hubs; they ensure the redistribution of forces and moments transmitted from the axle to the wheel [50]. The main share of failures occurs on the wheel of a railway car, due to the fact that the surface of the wheel of a railway car is the first to perceive dynamic loads. Let's take a closer look at some of the main malfunctions of wheel sets of railway cars.

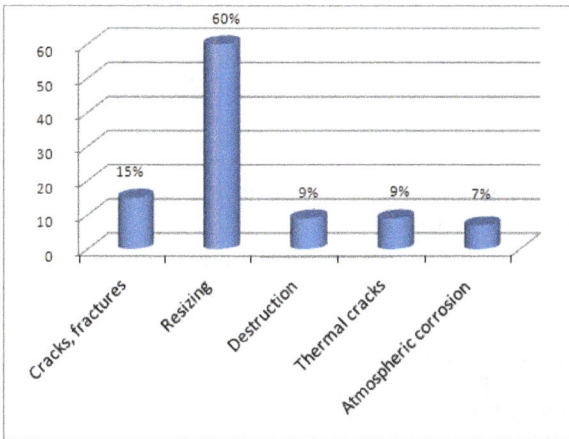

Figure 10. Distribution of railway wheel defects by type

Analyzing the histogram (Figure 10) it was found that the main share of defects relates to friction and wear, which is about 80%. This result is caused by the high tonnage of the car and high acceleration and deceleration speeds. Also, these processes are accompanied by intensive development of wear with a high coefficient of friction. The presence of friction forces creates high temperatures on the rolling surface, which increases stress inside the structure of steel wheels.

Circular uniform rolling (Figure 11, a) occurs when surface abrasion of the metal is worn out as a result of the action of contact friction forces during the movement modes of wheelsets: braking and slipping, while the hardness increases to *HB* 470 due to layering of the surface of the treadmill.

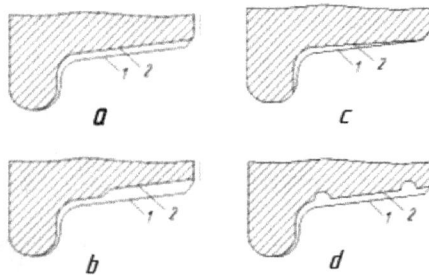

1 – unworn wheel profile; 2 – profile of the defective wheel

Figure 11. Defects in the wheel tread surface profile

Uneven stepped rolling (Figure 11, b) occurs due to uneven wear of the «treadmill-rail» contact patch in the direction of the rail head. Unevenness of rolled products is manifested due to uneven wear of the rolling surface, such as widening and crushing in the form of an influx of metal of the treadmill.

Wear, which is a consequence of friction with the rail during the winding movement of the rolling stock on straight sections of the track and when the car passes along radial curves, gives rise to a vertical undercut of the ridge (figure 1.11, c).

Ring grooves (Figure 11, d) or circular wear with local formation of depressions along the treadmill arise due to the difference in thermal conditions of interaction between the metal surface of the wheel and the composite block. Slider or local wear occurs when the wheel moves along the rail. When a carriage wheel moves curvilinearly along a rail in a skid or figure-eight pattern, deep deformation develops and hardening structures are formed, generating defects in the form of sliders. The formation of defects on the surface of the rim in the form of sliders, gouges, and welds can be detected by the characteristic knock of the wheel on the rails.

Changes in the original geometry of wheelsets - the distance between the inner edges of the wheels, reductions in the width and thickness of the wheel rim, are described by the discrepancy between the nominal permissible dimensions of the standard profile during repair, machining or operation.

Thus, due to the intensive operation of large-tonnage railway cars, the need to increase and maintain a high service life and reliability of wheels increases. To develop design and technological solutions to ensure the durability of wheels, it is necessary to study and justify the limits of permissible wear. An alternative to replacing with new wheels has been proposed to restore them; therefore, it is necessary to justify the defects that need to be restored and their surface properties. To make a reasonable choice of restoration method, it is also necessary to study the efficiency requirements for the maximum permissible wear of wheels.

1.4   Requirements for wheel sets regarding permissible wear of various types of defects

The wear of carriage wheels can be described as a complex multifactorial process in which the «wheel-rail» interaction occurs and the loss of metal from the working surface of the rim during operation occurs.

To increase the service life and reliability of rolling stock, the task is to develop new profiles of wheel pairs, ridge lubricants, and tribological lubricants that reduce wear, but do not eliminate the root causes of wear. Despite the reduction in railway track width from 1524 → 1520 mm and the introduction of a new type of rails P65 instead of P50, it is still necessary to solve the problem of creating and eliminating the maximum permissible wear of rolling stock wheel pairs and rails.

Such defective wear as a slider on a car wheel with a depth of 2 mm $<h>$1 mm is the limit for rolling the car to the nearest repair department to replace the worn gearbox without uncoupling the car from the rolling stock. Wheelsets with a rim widening of more than 5 mm and with existing spalling of the outer edge of the rim are recommended to be rolled out from under the cars [51].

If a weld of more than 2 mm is detected on wheel pairs during operation of the rolling stock, it must be brought to the nearest maintenance point (MTP). The fat is removed with an abrasive wheel. During long-term operation, as a result of metal fluidity (Figure 12), a significant influx of metal can form at the outer edge of the rolling surface, the chamfer of which is restored by turning on a lathe [52].

Figure 12. Result of wheel material yield

Cracks are divided according to the places of their formation (Figure 13-15). Some gouges develop along the tracks of crawlers, light spots and deposits. They appear due to structural changes in the metal, resulting from the formation of microcracks characteristic of a particularly bleached layer. Thedepth of suchgougesrarelyreaches 3 mm[53].

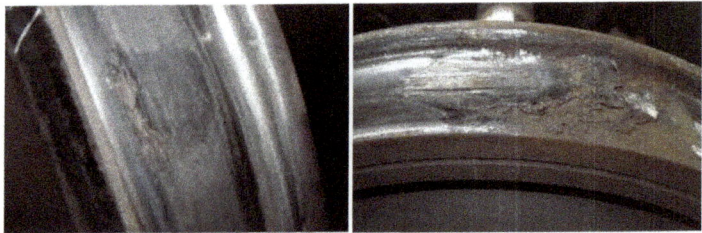

Figure 13. Shape of a double defect - gouges on the slider

Another type of gouge is the result of superficial fatigue failures, the development of shallow fatigue delaminations with resulting spalling of the metal edges. Fatigue cracks are formed under the influence of constant cyclically repeating contact loads. Inside fatigue gouges there are often cracks that go deep at an acute angle to the tread surface.

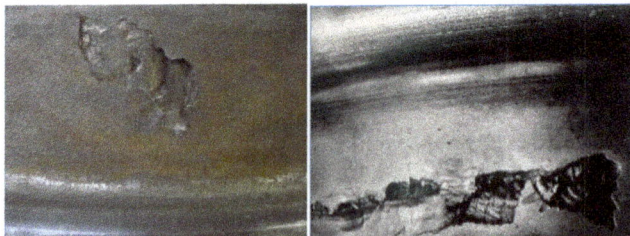

Figure 14. Result of the gouge at the site of fatigue crack formation

The next type of local gouges appears due to the spalling of transverse thermal cracks due to heating of the wheels during braking.

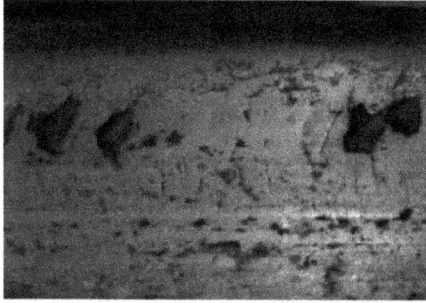

Figure 15. Thermal cracks in a hole

All types of dents can be eliminated by particularly careful turning of the wheels on a wheel lathe, but with large losses of the surface layer of metal. The permissible limits of gouges are depth $h \rightarrow 10$ mm, length $l \rightarrow 50$ mm, but the size limits of gouges on worn wheels reach a depth of 15-20 mm.

Thus, the requirements for wheel pairs regarding the wear limit of various types of defects are presented in Table 3; methods for eliminating damage and restoring the flange, rim and rolling surface of a carriage wheel are also shown, such as surface thermal hardening, optimization of the profile by turning and increasing wear resistance with tribological materials for increasing the service life of wheel pairs of rolling stock [54, 55].

Having analyzed the factors that cause intense wear of wheel pair units, methods for diagnosing the preliminary location of frequently occurring defects, their size and type of damage have been determined. The effectiveness and accuracy of diagnostic methods are determined by the conditions of maintenance, repair and operation. Having established the cause and effect relationship between wear and operational characteristics, it was possible to develop a table of the maximum wear of wheels subject to restoration and subsequent operation.

Table 3. Car wheel wear limits

Type of defect	Tolerance of wear size after restoration	Maximum permissible wear before restoration	Types of restoration effects
1	2	3	4
Rolling	uneven$\geq$ 2,0mm	- ultimate$h \leq 5,0$ mm - uneven$h \leq 0,7$ mm	turning, restoration
Max and min rim width; rim thickness	126…136 mm; rim thickness at least 35 mm;rimwidening$\geq$ 6,0 mm	nominal - 130 mm; broadening $\leq 3,0$ mm	turningordefect
Comb thickness wear	26…33$_{-0,5}$mm	33,0 мм	turning, surfacing
Vertical undercut of the ridge	$\geq$ 18 mm	height $\leq$ 18 mm	turning, grinding, heatstrengthening
Crawler	$h \leq 1$ mm	$h \leq 0,3$ mm	turning
Gouges	$h \geq 3$ mm, $l \geq 25$ mm	$h \leq 3$ mm, $l \leq 25$ mm	turning, welding, surfacing
Ring workings	$h \geq 1$ mm, $b \geq 15$ mm	$h \leq 1$mm, $b \leq 15$ mm	arcwelding, surfacing
Growth	$\leq 0,5$ mm	$h \leq 0,3$ mm	turning, heatstrengthening
d difference, ovality and eccentricity	$\leq 0,5$ mm	$\leq 1,0$ mm	turning and surfacing, heat strengthening
Distance between the inner edges of the wheels	1438…1433 mm, at different points up to 2 mm	1440^{+2}_{-1}mm	turningorchanging
Wheel profile	along the ridge $h \leq 1$ mm, Differences between treadmill surface and inner edge $\leq$0.5mm	standardsizes	grinding, turning
Roughness	$Rz\leq1,25$ (micrometer)	$Rz\leq80/Ra20$ (micrometer)	turning

Conclusion

From an analysis of the types of defects, it was established that the main share of failure occurs on the car wheel, namely on its treadmill and ridge. This is explained by the magnitude of the cyclically acting dynamic load and the impact moment. The most common type of defect is a change in the geometric dimensions and shape of the rolling surface of a railway car wheel, as well as microcracks and fractures.

It has been established that the reasons for changes in wheel geometry, the appearance of cracks and kinks are jamming of the wheel pair, metal defects, overloading of the wheel pair, violation of the requirements for assembling the wheel hub and axle in the quality of tolerances and fits, as well as uneven distribution of moments of force over the worn surface of the wheel during operation.

The main reason for the low service life of restored parts is that the process of thermal exposure is poorly controlled and leads to the problem of metal heating and mechanical processing of the rolling surface with a given accuracy and cleanliness of the machined surface. There is no single way to form the physical and mechanical properties of the surface and phase structure of a car wheel.

Modern methods and methods for increasing the service life of railway cars require timely detection of defects in wheel pairs, and the restoration of wheels is carried out by automatic submerged arc surfacing with a greater number of defects.

From the variety of defects studied, it was established that the presence of transverse cracks in any part of the wheel is categorically unacceptable. During operation of carriage wheels, uneven rolling of more than 2 mm, thickness of the flange and rim, as well as the distance between the inner edges of carriage wheels exceeding the established standards (GOST 10791-2004) are unacceptable. Wheel sets with uneven rolling of more than 2 mm are subject to rolling out and turning the contact profile.

Chapter 2. Studying the Effectiveness of Using Solid Rolled Wheel and Establishing Dependences of Changes in Dynamic Load on Technical and Operational Indicators of the Car

2.1 Analysis of the dynamics of changes in the service life of standard wheelsets

The dynamics of wear of the tread surface and wheel flanges increased 17-20 times higher than the underlying rolling stock operating technology; accordingly, this gave rise to abnormal metal losses of carriage wheels and flanges (approximately 70-80% of the total, compared to 20-30% of the technological output 70-80s).

The following methods for increasing service life are known, such as welding, surfacing, heat hardening of carriage wheels and flanges, lubrication, etc. [56-58]. Established methods extend the service life of wheels, but it is necessary to determine the root causes of increased wear directly arising from friction, slippage, corrosion, fatigue, overheating due to specific pressure, chemical destruction of metal in contact between the wheel and the rail [59-67].

The opinions of many researchers in the field of severe wear are not unanimous [68,69]. Past experience shows that despite significant changes in the reconstruction of the «car-track» system, such as changing the railway gauge from 1524 mm to 1520 mm, replacing plain bearings with rolling bearings with the loss of tribological lubricant and, as a consequence, an increase in friction forces, an increase in mass and the length of the rolling stock, as well as an increase in the load on the axle of the wheelset from 21 tons to 23 tons, the replacement of wooden beams (sleepers) with durable, strong reinforced concrete ones did not help reduce wear and tear to the present day [70-71].

The dynamics of fatigue wear intensity was studied by Professor I.V. Kragelsky. It has been established that the adhesion of the contact surface of the base metal and filler material in the transition zonemay be with a sliding friction coefficient of >0,22 under conditions of plastic contact and the presence of a lubricant lubricating film on the transition surface, >0,13 - under dry friction conditions [72].

The amount of wear directly depends on the service life of the rolling stock. Figure 16 shows the breakdown of mileage periods for rolling stock carriage wheels of various modifications. Figure 17 shows the distribution of car wheels depending on mileage. Under severe operating conditions, Figure 17, a - the average resource is determined to be ≈ 250 thousand km, with optimal operation Figure 17, b equal to ≈1 million km.

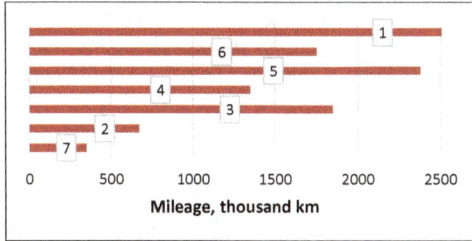

1 – high-speed electric trains;
2 – tilting trains;
3 – passenger trains;
4 – electric metro trains;
5 – heavy freight trains;
6 – standard freight cars;
7 – low-tonnage cars

Figure 16. Mileage of rolling stock of various modifications

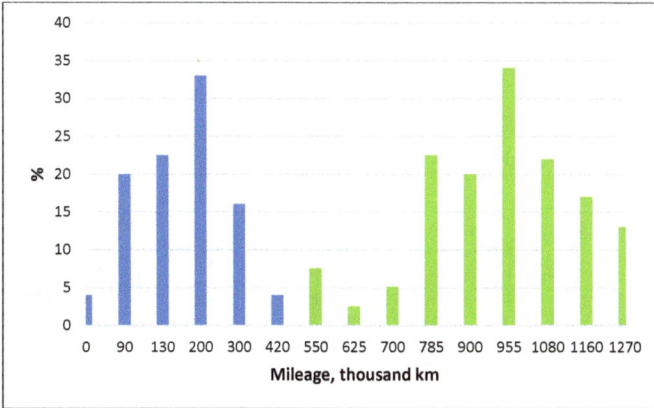

a   b
a – harsh operating conditions;
b – optimal operating conditions

Figure 17. Service life of freight car wheels

Consequently, all modifications of railway trains are similar in their performance, design and functionality, but when operating in different climatic zones they have different mileage and, accordingly, defective damage. To reduce the severity of the consequences of operational defects, it is necessary to constantly improve dynamic and structural changes in the life of a complex integrated system «body-bogie-wheel-rail»[73-77].

2.2 Assessing the strength and establishing the dependence of the change in the dynamic load of the wheel pair on the operational parameters of the car

The wheelset (WPS) is a significant unit of rolling stock, located between the car body and the rail and the first to receive all dynamic loads. To assess the strength and reliability of carriage wheel sets, it is necessary to determine all the loads acting on the gearbox and establish the dependencies of dynamic loads during operation using the calculation method.

The main forces and loads (Figure 18) falling on the wheel pair are determined using the following method [78].

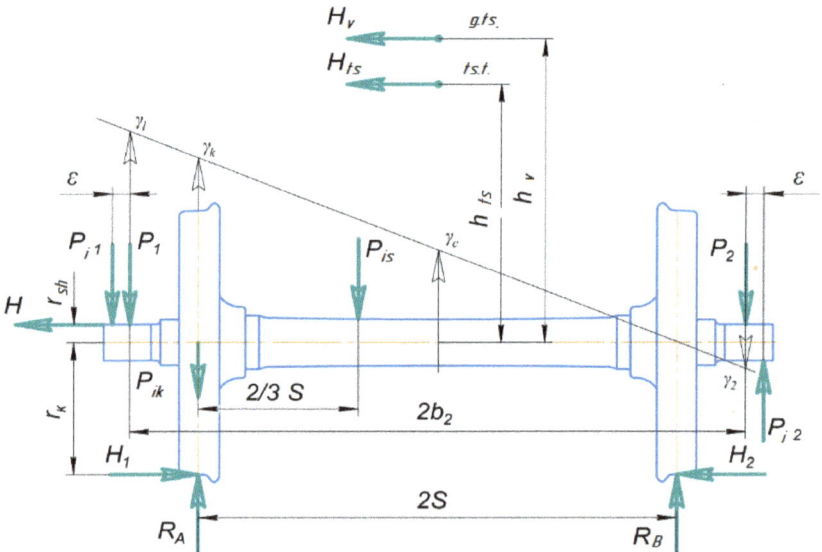

Figure 18. Scheme of application of power loads

The wear rate of a wheel pair is significantly influenced by the redistribution of moments of forces and structural gaps in the joints when the contact patch of the design tread surface profile deviates from the axial trajectory of movement.

The scientific hypothesis is that determining the boundaries of the deviation area of the contact patch of the wheel profile and the permissible coefficient of its deviation from the axial trajectory of movement will increase the overhaul life of the wheelset.

When studying the principles of dynamic load redistribution, changing the deviation of the contact patch of the wheel pair surface profile from the design axis of motion, it is necessary:

- determinepowerloads;

- determine the origin of stresses in gearbox elements;

- assess the strength of the wheel profile structure;

- justify the permissible limits of the area of deviation of the contact patch of the wheel profile from the design axis of movement.

The wheelset is under the influence of absolutely all force and weight loads acting on the car [79-81]. Let us determine the main forces that energy-efficiently influence the durability of a carriage wheel pair and are included in the strength calculation.

The action of the vertical static load P_{st} of a loaded car is found by the equality:

$$P_{st} = \frac{m_{br} - m_o m_{kp} + 2m_o m_{sh}}{2m_o} g \frac{1 + \bar{\lambda}}{2};$$

(11)

where m_{br}- the gross mass of the car;

m_0 – number of wheel pairs in the cart;

m_{kp} – mass of the gearbox;

m_{sh}– mass of the supporting part of the axle;

g – acceleration of gravity;

$\bar{\lambda}$ – average value of the wagon load capacity coefficienta.

Consequently, when solving equality (11), we include part of the weight of the wheel pair axle console in the car load, taking into account the incomplete use of the carrying capacity during the operation of the rolling stock.

Oscillations of the sprung masses of the wheel during the trajectory of the car create a vertical dynamic load, determined by the equality:

$$P_d = P_{st} \cdot k_d \tag{12}$$

The coefficient of vertical dynamics k_d is found by the equality:

$$k_d = \lambda_v \cdot \left(A + \frac{B_v}{f_{st}} \right); \tag{13}$$

where $\lambda_v = 1,0$ – trolley axle coefficient;

$A = 8,125$ – the value of determining the flexibility of the spring suspension of a car;

$B = 5,94 \cdot 10^4$ – value selected according to the type of car;

$f_{st} = 0,0463$ – static deflection of spring suspension, m;

$v = 15 \div 33$ – car speed, m/s.

In classical strength calculation methods, the main disadvantage is considered to be calculation under ideal conditions, in accordance with the manufacturer's regulatory standards. In real operating conditions, the contact rolling surfaces of wheel pairs wear out intensively [82-84]. Wear causes degradation of the contact surface according to its structural parameters (fatigue wear) and a change in spatial shape, design geometry and surface roughness $\sqrt{R_a} \leq \sqrt{R_a 2,5(\sqrt{\,})}$. For example, the vertical static load P_{st} on the wheel will change due to wear of the tread surface and the wheel flange. Changing the coefficient of utilization $\bar{\lambda}$ of the car's carrying capacity will affect the amplitude of the action of cyclic forces during operation. As a result, useful traction forces acting at a certain negative angle to the wear site form moments of resistance forces. Thus, it is necessary to solve the scientific problem of establishing the dependence of the static load P_{st} on the coefficient of utilization of the car's carrying capacity $\bar{\lambda}$, on the gross weight of the car m_{br} and also integrate classical calculations to real operating conditions by taking into account dynamically unstable loads. This problem was solved by establishing the dependences of the dynamic load P_d on the static load of the car R_{st} and the coefficient of vertical dynamics k_d on the speed of the car v. The regression equation of the studied dependencies is presented in Table 4.

Table 4. Regression equation of the studied dynamic load dependencies

Actualloadparameters	Regressionequations	Correlationcoefficient	Dependencies
P_{st}	$P_{st} = 8{,}430\lambda + 65{,}343$	$R^2 = 0{,}98$	dependence of the static load on the coefficient of utilization of the car's carrying capacity
P_{st}	$P_{st} = 0{,}023m^2_{br} - 0{,}770m_{br} - 5{,}573$	$R^2 = 0{,}99$	dependence of the static load on the weight of the loaded car
P_d	$P_d = 1{,}0862P_{st} - 65{,}6814$	$R^2 = 0{,}988$	dependence of the dynamic load on the static load of the car
K_d	$K_d = 0{,}0134\upsilon + 0{,}0297$	$R^2 = 0{,}988$	dependence of the coefficient of vertical dynamics on the speed of the car

The rolling of the wheels occurs in asymmetrical vibrations, as a result of which the vertical dynamic load is taken from one journal and the load from the centrifugal force (Figure 19) is represented by finding the centrifugal force H_{ts} of the car of one gearbox from the height of the center of mass h_{ts} of the car from the gearbox axis and the distance between the middle of the journals of the gearbox axis $2b_2$ by equality:

$$P_{ts} = H_{ts}h_{ts}/2b_2. \tag{14}$$

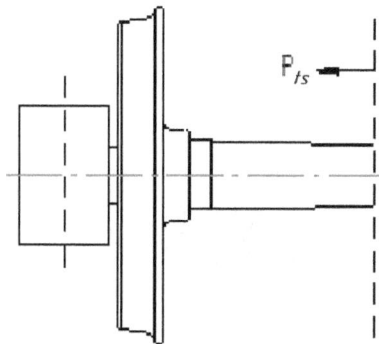

Figure 19. Load from centrifugal force

Next, we find the dynamic load from wind pressure (Figure 20), acting on the side surface of the car by the equality:

$$P_v = \omega F \frac{h_v}{2_{b_2} \cdot m_o};$$

(15)

where $\omega F = H_v$ – the effect of wind pressure on the car (Figure 21);

h_v – distance from the wind pressure equinox to the WPS axis, m;

ω - wind pressure directed perpendicular to the side of the car;

F - side surface area of the car, m².

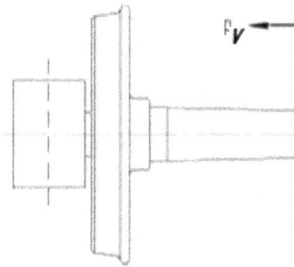

Figure 20. Determination of vertical load from wind pressure

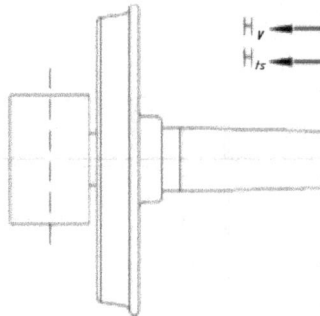

Figure 21. Effect of wind pressure on the car and centrifugal force

These forces are taken into account as a static load due to a slow change in time and are taken as $P_{ts} = P_v = 1$. We determine the total vertical load (Figure 22):

on the left neck

$$P_1 = P_{st}(1 + k_d) + P_{ts} + P_v, \qquad (16)$$

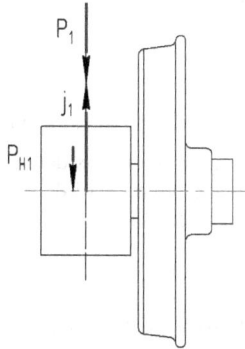

Figure 22. Total vertical load acting on the left neck

on the right neck (Figure 23)

$$P_2 = P_{st} - P_{ts} - P_v. \qquad (17)$$

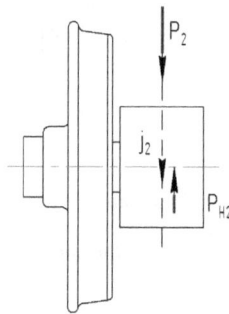

Figure 23. Scheme for determining the total vertical load acting on the right neck

Let us imagine the vertical load of the inertia forces of unsprung masses acting on the left journal of the axle:

$$P_{H1} = m_1 \cdot j_1; \tag{18}$$

where m_1 - sum of masses of unsprung parts, t.

We determine the acceleration of the left axle box unit from the equality:

$$j_1 = \frac{204 + DV}{\sqrt{m_H}} \cdot g, \tag{19}$$

where D – coefficient determined by the type of car and the speed of movement.

The mass of unsprung parts resting on the rail is determined by adding the mass of the axle box, the mass of the gearbox and the mass of the action on the axle box of the bogie side frame units and the spring set, by equality

$$m_H = 0{,}5m_{kp} + m_b + 0{,}5(m_{rb} + m_{pk}). \tag{20}$$

Using a similar method, calculations were made for the right neck of the wheel (Figure 24).

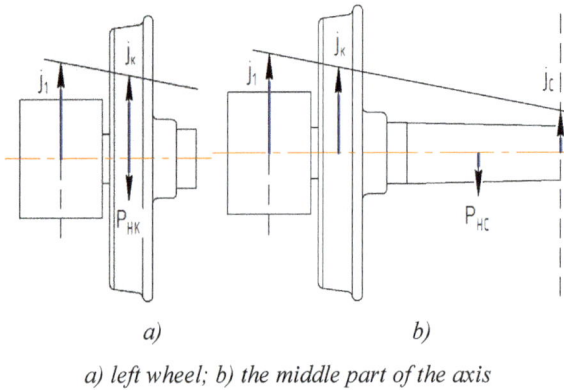

a) left wheel; b) the middle part of the axis

Figure 24. Calculation scheme for determining the inertia force

The friction force in the contact «second wheel – rail» is determined by the following equality (Figure 25):

$$H_2 = \mu N_v, \tag{21}$$

where μ – lateral sliding friction coefficient, $\mu = 0{,}25$;

     N_v - vertical reaction in the axle support.

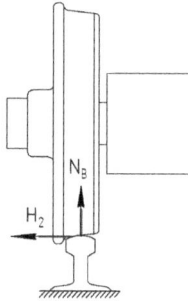

Figure 25. Friction force of the «wheel-rail» contact

Vertical reactions in the axle supports are determined as follows:

- for left axle support

$$N_{\text{H}} = P_1 \frac{l_2+2s}{2s} + P_{H1} \frac{l_5+l_2+2s}{2s} + H \frac{r+r_1}{2s} + P_{HK} + \frac{2}{3}P_{HC} - P_2 \frac{l_2}{2s} + P_{H2} \frac{l_2+l_5}{2s}, \tag{22}$$

where H – frame strength;

 $2s$ – distance between WPS running tracks;

 l_2 – distance from the center of the axle journal to the plane of the wheel tread surface;

 l_5 – eccentricity relative to the center of the axle neck;

 r – car wheel radius;

 r_1 – wheel pair axle journal radius.

- for right axle support

$$N_{\text{B}} = P_2 \frac{l_2+2s}{2s} - P_{H2} \frac{l_4+l_2+2s}{2s} - H \frac{r+r_1}{2s} + \frac{1}{3}P_{HC} - P_1 \frac{l_2}{2s} + P_{H1} \frac{l_2+l_4}{2s}, \tag{23}$$

where l_4 – distance from the center of the axle journal to the center of the axle.

We find the frame force H from the equality:

$$H = \frac{m_{br}}{m_0} g \cdot k_{dg}. \tag{24}$$

The coefficient of vertical dynamics is determined by the equality (25):

$$k_{dg} = \bar{k}_{\text{дг}} \cdot \sqrt[4]{\frac{4}{\pi} \cdot ln \frac{1}{1-P(k_{dg})}}, \tag{25}$$

where $P(k_{dg})$ – confidence probability equal to 0,97.

The average value k_{dg} is determined by the equality:

$$\bar{k}_{dg} = B \cdot \text{Б}(5 + V), \tag{26}$$

where B -influence coefficient of the number of axles n in the cart;

Б – coefficient selected depending on the type of carriage chassis (Б=0,003).

$$B = \frac{n+2}{2n}. \tag{27}$$

Consequently, the force applied to the flange of the wheel H_1 is determined by the equality (Figure 26):

$$H_1 = H + H_2 \tag{28}$$

$$H_1 = 41,4 + 3,67 = 45,07 \ \kappa N.$$

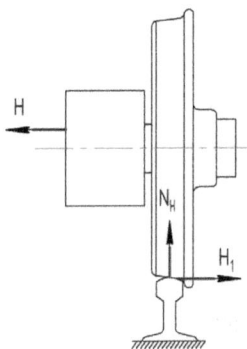

Figure 26. Vertical reaction of the left axle support

Thus, when carrying out a strength calculation as a result of the action of all force and external loads, the especially loaded state of the wheel rim, flange and tread circle was determined(P_{st}=109,5 кN, P_d=32,8 кN, P_{ts}=11,5 кN, P_v=0,25 кN, H_{ts}=17,8 кN, H_v=1200 Pa, P_1=166,7 кN, P_2=105,8 кN, P_{H1}=187 кN, P_{H2}=27,1 кN, P_{HK}=49,7 кN, P_{HC}=12,72 кN, H=41,4 кN, H_2=3,67 кN, H_1=45,07 кN). It has been established by calculation that when moving along curved and straight sections of the rail track, wheel slippage occurs, leading to thermal and mechanical failures of the tread surface and, as a consequence, to rapid chipping of areas where contact stresses are applied[85,86].

It has been established that the least studied element in wheel wear is the contact patch. It is always irregular in shape and difficult to measure. It has also been established that during operation, due to imperfections in the design and differences in wheel materials, the contact patch is constantly formed in different areas of the treadmill and ridge, and research has proven that progressive wear of the wheel and ridge occurs when the contact patch is formed with a deviation from the design axis of symmetry. Since the irregular shape of the contact patch generates additional resistance forces, moments of inertia and cyclically unstable shock loads, which leads to progressive wear not only of the wheel, but also of the entire part of the trolley. The physical meaning of defect generation at the structural level has been proven.

Wear at the point of contact occurs as a result of fatigue stresses in the wheel. The obtained calculations made it possible to determine the influence of static and dynamic loads on the formation of the deviation area of the contact patch of the wheel profile, taking into account the different load-carrying capacity of the car and its speed: the dependence of the static load on the coefficient of utilization of the car's carrying capacity (P_{st}=8,430λ+65,34); dependence of the dynamic load on the static load of the car (P_d= 1,0862$P_{cт}$-65,681); dependence of the coefficient of vertical dynamics on the speed of the car (k_d = 0,013v + 0,0297) and the dependence of the static load on the gross weight of the car (P_{st} = 0,02m^2_{br} – 0,770m_{br} – 5,573).

In this regard, there is an urgent task to substantiate the main factors influencing wear.

2.3 Development of a methodology for substantiating the main factors influencing wheel wear

The efficiency and safety of operation of a freight car is significantly influenced by the technical condition of the wheelset, platform and the process of interaction of the contact rolling surface with the rail, which is accompanied by large dynamic loads. Researchers Buinusov A.P., Glazunov D.V., Myamlin S.V., PanasenkoV.Ya., Dumpala R, Chandran M., Rao M.S.R. found that solid-rolled wheels are predominant in terms of operation compared to banded ones [98]. During the long-term movement of rolling stock, the contact of the wheel with the rail is subjected to the progressive influence of dynamic loads (P_d, P_{ts}), not constant in time. Despite the small contact area between the wheel and the rail, the wheels still bear large static and dynamic loads with a force of ≤ 110 kN. From the impact of these force loads in the contact zone of the wheel with the rail with increased fatigue stresses $\sigma_к$ and deformations, dangerous surface failures and destruction appear.

The dynamic impact of a changing complex cyclic-oscillatory load directly depends on the intensity of changes in the design profile of the wheel (wear - defect), operating modes (acceleration - braking - movement) and the sum of moments of forces arising during different periods of wheel contact. The durability of a wheel pair of a railway car is mainly determined by its dependence on the dynamic and strength characteristics of the wheel, as well as on the design features of the freight trolley (Figure 27).

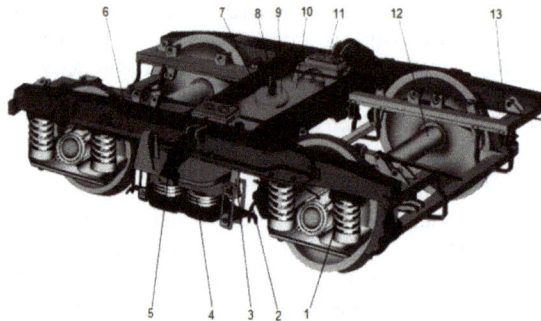

1 - axle-box suspension; 2 – bracket; 3 – elongated earring;
4 – central suspension; 5 - hydraulic vibration dampers;
6 - longitudinal leads; 7 – frame; 8 – kingpin; 9 – thrust bearing;
10 - bolster; 11 – sliders; 12 - two wheelsets;
13 - brake lever transmission

Figure 27. Design of a carriage bogie

Experiencing significant cyclic, statistical, shock and dynamic loads, the elements of the wheel pair wear out intensively. As a result of wear, defects are formed that have a wide range of structural and external changes. The presence of the previously discussed defects in the wheel pairs of railway cars reduces the efficiency of their operation, and in some cases they are dangerous and require repair and restoration work. Typical sources of occurrence of various wheel failures in different interaction zones and their origin are shown in Figure 28.

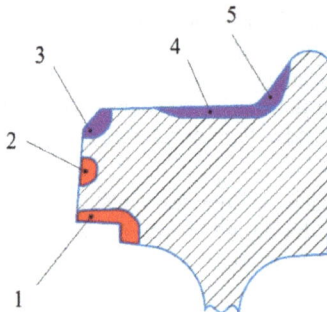

1, 2 – failures of thermal origin;
3, 4, 5 – failures of mechanical origin

Figure 28. Pronounced area of the image of defects in a railway wheel

The next important factor in the formation of wear of the «wheel-flange-rail» contact is the transition of wear to the flange area. The wheel flange helps the wheel to roll steadily along the rails; it maintains the stability of the cart, i.e. prevents the wheel from leaving the rail. It has been established that the optimal value of the angle of inclination of the flange generatrix to the wheel tread reaches a value of $\approx 66°$ during the trajectory of movement in curved sections after the turning operation. This optimal value of the ridge inclination angle reduces the contact pressures in the area of contact of the two zones and thereby reducing the wear of the two surfaces.

The hardness of wagon wheel sets is one of the main factors of intense wear. It has been determined that the hardness on the rolling surface should be higher than the hardness on the ridge and the difference in their hardness is approximately 90-100 HB. As the hardness of a car wheel increases, its service life increases several times; it is also necessary to take into account the method of thermal restoration of the wheel surface to obtain optimal hardness and improve the quality of the wheels.

Currently, there is no systematic approach to substantiating the reasons for the formation of such a heterogeneous range of defects. In their research, scientists Myamlin S.V. and Rao M.S.R. ambiguously interpret the physical meaning of the process of defect formation. Classical analysis and strength calculations of a wheel are reduced to the study of the acting forces applied to an equilibrium system, taking into account the static characteristics of the wheel dimensions and wheel material. The defect is examined as a consequence, the result of fatigue wear (cracks, tears, destruction, spalling, breaks, sagging, etc.), but the original cause of its formation is not investigated. However, this calculation method does not take into account the acting moments of forces at different time intervals in different operating modes, as well as the influence of changes in the wheel profile and the deviation of the contact spot from the axial trajectory of movement.

Many methods for the wear of carriage wheels are based on the fact that the amount of worn wheel metal and the friction force in the "wheel-rail" contact are directly proportional [87, 88]. The most effective and objective wear method is Archard's theory [89-91]. This technique is based on finding the transition point by calculation between low and deep wear. Incorrect determination of the position of this transition and the absence of a wear coefficient makes the problem of determining the mathematical parameters of wheel wear during operation relevant.

This problem is solved by finding the ratio of the power of friction forces at the transition point of the «wheel-rail» contact to the area of the contact patch using the method of mathematical modeling in the MATLAB software package when moving on straight and curved sections of the track. At the same time, increased wear intensity is observed on curved sections $R \leq 600$ m.

The power of friction forces in the «wheel-rail» contact is determined by the equality:

$$P = \frac{F_i \cdot v_0}{F},\tag{29}$$

where v_0– rolling stock speed, m/s;

F–friction force in the «wheel-rail» contact,m^2;

F_i–wear factor,H.

Taking into account all the factors affecting the wear of the «wheel-flange-rail» system, it is necessary to express them through the integral index F_i. In this case, the increase in wheel wear, taking into account driving modes with different time intervals and material hardness, will be determined using the proposed equality:

$$F_i = \int_{t_i}^{t_1} \frac{fNW}{G\sigma_\text{н}},$$

(30)

where f–friction coefficient;

N– normal pressure at the point of contact between the wheel flange and the rail;

W–sliding of the wheel flange along the rail;

G– contact area between wheel flange and rail,

$\sigma_\text{н}$ – contact stress in the wheel, characterizing the hardness of the material.

The action of normal pressure at the transition point of contact «wheel-rail ridge» can be expressed by the equality:

$$N = Pf / \sin \gamma,$$

(31)

where γ– angle of inclination of the working edge of the ridge to the horizon axis of the wheelset.

Passing along the trajectory of curved sections of rolling stock, there is a lack of rolling friction, which means sliding along the rails due to the difference in the diametrical turning curves of the track and the difference in the distances between the rails[92-93].

Accordingly, the total relative sliding of the wheel flange along the rail is determined by the equality:

$$W = \sqrt{\left(\frac{S}{2R} - \frac{a}{R}\right)^2 + \left(\frac{x_1}{R\cos\gamma}\right)^2 + \left(\frac{a}{r}\frac{x_1}{R\cos\gamma}\right)^2},$$

(32)

where S–distance between rail axes, m;

x_1 – distance from the turning pole to the geometric axis of the first wheelset, m;

R – curve radius, m;

a – depth of contact between the wheel flange and the rail head, m;

r– wheel radius, m.

The contact area of the wheel and rail is determined by the ratio:

$$G = 1 + 30\frac{x_1}{R}. \tag{33}$$

The expressed formula for finding the slip value expresses the influence of the factors of the "wheel-rail" system on the intensity of their wear. Therefore, we describe the phenomenon of slipping by the coefficient obtained by S. M. Akurlevsky[94-95]:

$$\delta = \frac{F_k}{4.1\sqrt{1 - F_k}} \tag{34}$$

The ratio of the circumferential force on the wheel rim to its limiting value for adhesion is determined by the equality:

$$F_k = \frac{F_{\text{окр}}}{F_c}. $$

Transforming formula (32) taking into account the wear factor and δ, we obtain

$$W_D = \sqrt{(\frac{S}{2R} - (1 + \delta)\frac{a}{r}\Phi_\text{н})^2 + \left(\frac{x_1}{R\cos\gamma}\right)^2 + (1 + \delta)\frac{a}{r}\frac{x_1}{R\cos\gamma})^2}. \tag{35}$$

Consequently, the system of equations for the power of friction forces in the contact patch of the wheel with the rail and the total relative slip will take the final form

$$\begin{cases} P = \int_{t_i}^{t_1} \frac{fNW}{G\sigma_\text{н}} \frac{v_0}{F}; \\ W = \sqrt{(\frac{S}{2R} - (1 + \delta)\frac{a}{r}\Phi_\text{н})^2 + \left(\frac{x_1}{R\cos\gamma}\right)^2 + (1 + \delta)\frac{a}{r}\frac{x_1}{R\cos\gamma})^2}. \end{cases} \tag{36}$$

It has been established that the dynamic impact of a changing complex cyclic-oscillatory load directly depends on the intensity of changes in the design profile of the wheel, operating modes (acceleration - steady motion - braking) and the sum of moments of forces arising during different periods of wheel contact.

Thus, using the method of mathematical modeling of the movement of rolling stock in straight and curved sections of the track, the value of the average power of the friction force in the contact of wheels with rails (7 MW/m^2), related to the area of the contact patch in the transition zone from a low stage of wear (7.2- 20.2 W/m^2) to enhanced (18-20,2 W/m^2) in the time interval, including starting, acceleration, steady movement. The wheel wear model is based on finding wear coefficients for two stages and friction coefficients on the flange and tread surface of the wheel. The intensity of wear of a car wheel depends on the contact patch between the rolling surface and the rail, and from an increase in the friction and sliding force during the movement of the car, the contact patch moves towards the wheel flange; as a result of this phenomenon, wear resistance decreases and the proposed coefficients must be taken into account.

By studying the interaction of the wheel with the rail, it is substantiated that the power of the friction force in the contact patch «wheel-rail» and the total relative slip W, the angle of approach of the wheel on the rail, the contact area of the wheel flange and the rail, the depth of contact of the wheel flange and the rail head a are the main factors, affecting wheel wear.

2.4 Improvement of the mathematical model of wheel wear taking into account dynamic load and coefficient of friction in the flange contact patch

Intensive wear leads to frequent derailment of rolling stock due to increased failures of the wheel-rail system. The level of operational safety of a rolling stock as a mechanical multi-level system is determined by its resistance to derailment.

Rolling of rolling stock under different speed loads leads to the appearance of additional lateral forces from the interaction of the wheel flange with the rail. With an increase in the speed of movement of a railway car and passage along curved tracks, the instability of the carriage increases and there is a huge risk of derailment, and as a result, an accident and loss of life. Thus, to solve the scientific problem, it is necessary to conduct research on the development of such units and elements of rolling stock that minimize the failure of rolling stock to derail due to abrasion of the wheel flange towards the rail head. Scientific research is aimed at developing modern design solutions to determine the optimal stability parameters of rolling stock [96,97].

We will determine the mathematical model of rolling stock stability by the force loading action of wheels and rails when rolling the train along curved sections of the track [98-99].

The friction forces arising between the wheels and rails complicate the movement of the carriage bogie during lateral displacement and on turns as a result of the complex oscillatory motion of the car: $P\mu$ - friction force at each point of contact (μ - sliding friction coefficient), P - static load acting from the wheel on the rail. The components of these forces – longitudinal Hi and transverse Vi – are determined analytically (i is the WPS number) [100-101]:

$$H_i = P\mu \frac{s}{\sqrt{x_i^2 + s^2}} = P\mu \frac{0{,}5l_k}{\sqrt{x_i^2 + \frac{l_k^2}{4}}} \; ; \tag{37}$$

$$V_i = P\mu \frac{x_i}{\sqrt{x_i^2 + s^2}} = P\mu \frac{x_i}{\sqrt{x_i^2 + \frac{l_k^2}{4}}} \; ; \tag{38}$$

The effective static load on the rail is found from the equality:

$$P = P_p + q, \tag{39}$$

where P_p - load from the sprung weight as the wheel rolls along curves;

q- unsprung weight.

The system of equations for the equilibrium stability of a carriage bogie and the sum of moments is represented by the equality:

$$\begin{cases} -y_1 + C_T + \sum V_i = 0 \\ C_T x_2 + \sum M_{TP} - M_V = 0. \end{cases} \tag{40}$$

According to this system of equations, the carriage bogie, moving along curved tracks and sharp turns, perceives the action of the centrifugal force of inertia $C_T$and the guiding force from the outer rail y_l:

$$C_T = 2mP \left(\frac{v^2}{gR} - \frac{h}{2s} \right), \tag{41}$$

where h – outer rail elevation,

$$h = \frac{2sv^2}{gR} = 12{,}5\frac{v^2}{R};$$

where$m$–number of axles in the cart;

v–rolling stock speed;

$2s$–the distance between the surfaces of the wheelset treadmills;

R–unstable path curve radius.

Let us replace the centrifugal force of inertia with the unsuppressed acceleration defined by the equality:

$$a_{nu} = \frac{v^2}{R} - \frac{h}{2s} g.$$

The total moment of friction forces at the point where the left wheel hits the rail along the trajectory of the front WPS is expressed by the equality:

$$\sum M_{TP} = 2V_2 l_2 - 2V_3 l_3 - (H_1 + H_2 + H_3) l_k,$$

where l_k – distance between ski centers WPS;

$l_2$andl_3 – distance between the axes of adjacent WPSs.

The pole distance in the range from L to l_T is found:

$$x_2 = x_1 - l_T, here x_1 = \frac{L}{2} + \frac{R(u_b - u_a)}{L};$$

where u_a – the amount of displacement of the front WPS when the flange of the left wheel runs onto the outer rail;

u_b – the gap between the flange of the left wheel and the head of the outer rail. It is necessary to take into account the restoring moment M_v:

$$M_v = \left(P_{shH} + P_i'\right)(l_1 - l_2), \tag{42}$$

where $P_{shH} = 1,3 P_{shst} = 1,3 \, (Pst - q)$ – static force on the axle journal;

$P_i' = \frac{G_b \bar{\omega}}{g}$ – inertia force of axle box and balancer: G_b – weight of axle box and balancer;

$\bar{\omega} = \left(2 + 0,13 \frac{v}{\sqrt[3]{(2q)^2}}\right)$ – circular frequency of reference and wobble.

The system of equilibrium equations for the carriage bogie will take its final form:

$$
\begin{cases}
-Y_1 + 2m(P_p + q)\left(\frac{v^2}{gR} - \frac{h}{2s}\right) + \\
\sum(P_p + q)\mu \dfrac{x_1}{\sqrt{x_i^2 + \frac{l_k^2}{4}}} = 0; \\
2m(P_p + q)\left(\frac{v^2}{gR} - \frac{h}{2s}\right) \cdot \left(\frac{L}{2} + \frac{R(u_b - u_a)}{L} - l_T\right) + \\
+(2V_2 l_2 - 2V_3 l_3 - (H_1 + H_2 + H_3)l_k) - \\
-\left(1,3(P_{st} - q) + \frac{G_b}{g}\left(2 + 0,13 \frac{v}{\sqrt[3]{(2q)^2}}\right)\right)(l_1 - l_2) = 0.
\end{cases} \tag{43}
$$

We will determine the stability assessment according to the condition of the safety factor of the stability of the carriage wheel against derailment:

$$k_u = \frac{tg\beta - \mu}{1 + \mu tg\beta}\left(\frac{P_B}{P_b}\right)\kappa \frac{x_1}{R\cos\gamma}\frac{FG}{\sigma_{Mx}} > 1, \tag{44}$$

where β–angle of inclination of the cone-forming surface of the wheel flange with the horizontal, β=60...65°;

$G and F$ – area and friction force;

σ_{Mx}- contact stress;

μ–slip coefficient (friction of «wheel-rail» surfaces).

The value $\frac{tg\beta-\mu}{1+\mu tg\beta}$ is unchanged, the ratio$\left(\frac{P_B}{P_b}\right)$ is predominant; to satisfy condition (44), it is necessary to take into account the condition of the forces $\left(P_v > P_b \Rightarrow \frac{P_v}{P_b} > 1\right)$. Theforcesrepresentedareshownin Figure 29.

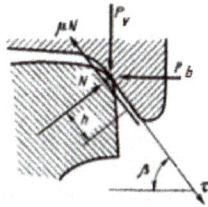

Figure 29. Image of the forces between the wheel flange and the side face of the rail head

Figure 30 shows a diagram of the action of forces on the gearbox on curved sections of the track: H_1 – lateral pressure on the guide wheel; H_2 – friction force distributed between the wheel rim and the inner rail, Q_{shsr}– static load on the axle journal.

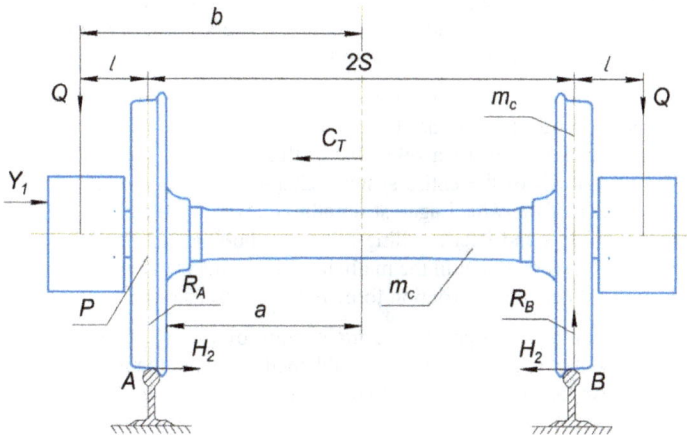

Figure 30. Representation of forces on path curves

Consequently, the equality for determining the safety factor of a wheel against derailment takes the form:

$$
k_y = \frac{tg\beta - \mu}{1 + \mu tg\beta}\left(\frac{P_e}{P_b}\right)\kappa\frac{x_1}{R\cos\gamma}\frac{FG}{\sigma_{Mx}} = \frac{tg\beta - \mu}{tg\beta - \mu}\times
$$

$$
(2Q_{shst}\left[\frac{b - a_{1,2}}{l}(1 - k_{\text{дв}}) \pm \frac{b}{l}k_{dbk}\right] \pm
$$

$$
\left(\pm F_p\frac{r}{l} + \frac{q_{kp}\frac{b-a_{1,2}}{l}}{F_p+\mu P_2}\right)\kappa\frac{x_1}{R\cos\gamma}\frac{FG}{\sigma_{Mx}}, \tag{45}
$$

where $q_{kp}=m_{\text{kp}}g$–wheelset weight;

$k_{dv} = \frac{Q_{dsh1}+Q_{dsh2}}{2Q_{\text{шст}}}$ – coefficient of dynamics of vertical vibrations of the car body;

$k_{dbk} = \frac{Q_{dsh1}-Q_{dsh2}}{2Q_{shst}}$ – coefficient of dynamics of lateral rolling of the car body.

In its presented form, the safety factor of a wheel against derailment more fully takes into account the distribution of power loads and stresses acting in contact and more objectively and accurately assesses traffic safety conditions.

Thus, the mathematical description of the dynamic force interaction of wheels, flange and rails has been improved, and a description has been made of the safety indicators of rolling stock, including the friction coefficient at the flange contact patch, the safety factor against derailment, which influence the formation of the contact patch and the formed contact stresses between the ridge, the treadmill and the rail. There is an integration of normal and generated contact stresses, friction forces and relative slipping at the site of origin of the contact spot, due to the action of a force load.

The mathematical model of wheel wear has been improved by introducing indicators and factors influencing progressive wear. The novelty of the mathematical model lies in the description and establishment of a relationship that describes the relationship between wear and dynamic loads of the entire system, characterized by the safety factor for the equilibrium stability of the wheel against derailment, the coefficient of dynamics due to vertical oscillations k_{dv} and lateral rolling of the car body k_{dbk}. The mechanical change in the properties of a worn wheel in the mathematical model is taken into account through the relationship between area G, friction force F and contact stress σ_{Mx}.

Consequently, a complex system of the main units of rolling stock operation can be represented by taking into account the established strength loads and coefficients of friction and stability at the origin of the contact patch of the wheel and flange with the rail.

Wear limits in the form of rolled products of more than 2 mm are subject to restoration by laser spraying with a powder of a multicomponent composition, and faults exceeding 2 mm, such as gouges, ridge defects, and rim widening, are restored by laser surfacing.

Conclusion

The problem of increasing the service life, ensuring the durability and efficiency of wheels of rolling stock cars is solved by developing new technological processes and restoration methods to repeatedly extend the life cycle of worn wheels, aimed at increasing the wear resistance of car wheels of the railway system, considered as a single whole car with a rail track.

The problem of assessing the adequacy of calculations has been solved and the permissible limits of internal wheel stresses arising on the contact surface of the wheel under the action of cyclically changing dynamic loads have been established. Wear at the point of contact occurs as a result of fatigue stresses in the wheel. The influence of static and dynamic loads on the formation of the deviation area of the contact patch of the wheel profile is substantiated, taking into account the different load capacity of the car and its speed.

According to the results of an improved mathematical model of wear of the tread surface and ridge, in addition to the dynamic forces of action when moving on straight and curved sections, it is necessary to take into account the coefficients of friction, slippage, and the safety factor against derailment, which cyclically act on the wheel and move the contact patch towards the ridge wheels. It has been determined that the average value of friction force power $P \leq 7$ MW/m^2 is the same on the rolling surface on straight sections of the track and depends on the area of the contact patch. In curved sections of the track and on sharp turns, the power on the tread surface of the oncoming wheel is higher for the non-oncoming wheel: for curves of large and medium radii - 7.2-20.2 W/m^2 compared to 4-6 W/m^2 and for curves of small radius - 18-20.2 W/m^2 compared to 6.5-9.5 W/m^2. The optimal angle of inclination of the wagon wheel flanges is 66°.

The wheel wear model is based on the choice of wear coefficients for low and high degrees of wear and friction coefficients on the flange and tread surface of the wheel. The intensity of wear of a car wheel depends on the contact patch between the rolling surface and the rail, and on the increase in the force of friction and sliding; during the movement of the car, the contact patch moves towards the wheel flange, as a result of this phenomenon, wear resistance decreases and the proposed coefficients must be taken into account.

Chapter 3. Research of Methods for Increasing the Durability of Wheel Pairs of Railway Cars

3.1   Study of effective methods for restoring worn surfaces of dynamic systems of railway cars

Many effective methods for restoring fatigued carriage wheel sets have been implemented and studied in the repair and restoration industry. All known restoration methods must satisfy the technology and quality indicators of the technological, economic and structured accident-free long-term reliable life cycle of the railway system under different operating conditions. When selecting a method for restoring wheel sets, one should proceed from the minimum cost of restoration and ensuring the service life of the restored wheel is at the level of a new one.

In the practice of car building, frequent failures in the wheel rim and flange during the operation of rolling stock are eliminated using mechanical grinding or thermal effects [102].

On the railways of the CIS, wheels that are restored by repeated repair or turning (Figure 31) are widely used due to such a design feature as thickening of the rim. But the cost of these wheels is much higher than wheels with thinner rims of single or double turning.

Figure 31. Thick wheels with multiple repairs

From domestic and foreign practice, the following methods are known for restoring the profile of wheels with and without rolling them out from under the car according to the processing principle: restoration by copying, restoration by cutting and combined processing. A promising method of restoration by copying (Figure 32) is the method using computer numerical control (CNC) machines, in which you set a given parameter for the movement of the cutting tool, written in the program. Restoring wheels by turning or cutting occurs by processing the profile with a shaped cutting tool. In this case, the profile of the cutting tool is chosen to be the opposite of the wheel profile. Combined profile processing integrates the copying method and turning with a cutting tool. The difference between this method is the possibility of additional supply of electrical, thermal or chemical energy to the processing zone before, during or after the turning process with preliminary thermal heating of the wheel tread surface.

Figure 32. Restoring the wheel profile using a copy

Significant disadvantages of the studied methods revealed rapid wear of the tool; a fairly high roughness of the resulting surface, depending on the design of the shaped tool; complex and highly qualified equipment maintenance, sometimes impossible in the conditions of repair enterprises; turning reduces the thickness and shortens the life of the wheel.

The most energy-efficient way to increase the durability and wear resistance of wheels is surfacing the surface of worn-out (Figure 33) carriage wheels [103]. Surfacing compared to turning and restoring using a carbon copy has obvious advantages: surfacing of car wheels reduces the technological wear of the rim until its resource is completely exhausted; idle time of rolling stock for the necessary intermediate turning is eliminated; high productivity of the surfacing process; no highly qualified repairman required; possibility of carrying out surfacing works in hard-to-reach remote places[104,105].

Figure 33. Surfacing of worn tread and flange surfaces

At railway transport enterprises for special repair purposes, more than forty methods of welding, surfacing and spraying are used (Instruction TsT-336) [106]. Modern progress in the development of technologies and methods for restoring and repairing worn-out carriage wheels by surfacing and spraying does not stand still. Restoration by welding, surfacing and spraying using technological means entails the development and introduction of resource-saving technologies into production.

The disadvantages of surfacing are uneven heating of the surface; obtaining a poor-quality deposited layer due to large penetration of the substrate metal and, as a consequence, a decrease in wear resistance; preheating of the surface to be deposited, which increases the cost of the restoration process; large allowance for finishing fine mechanical turning and significant heating of the welded part, which deforms the shape of the part after cooling.

Traditional methods for restoring the rim and flange of a wheel are electric arc or electroslag surfacing of filler material on the worn part of the wheel pair [107]. Electric arc surfacing is used to restore mainly the surfaces of small-sized parts. Restoring the wheel pair rim to its original geometric parameters with this type of surfacing undoubtedly increases the service life, but this method of hardening leads to the appearance of microcracks as a result of strong thermal heating of the fusion zone. To obtain a high-quality deposited layer by electric arc surfacing, a filler material is used that contains both alloying elements and protection from external environmental influences. To obtain a deposited layer with improved physical and chemical properties, choose wire or powder with a high content of carbon and alloying elements, and use flux as protection [108].

A significant disadvantage of this method of restoring the rim and flange of a rolling stock wheel is the increased plasticity of the deposited layer, which, under further harsh operating conditions, leads to repeated destruction, fragility and fragility of the deposited surface. Therefore, although this method of restoration is currently used at repair and restoration enterprises, it has not found wide application and does not limit the way to solve the problem of developing or improving an energy-efficient method of restoration. Table 5 presents some methods for restoring carriage wheels using mechanical and thermal methods.

Table 5. Technical characteristics of wheel restoration methods

Recoverymethod	Specifications	Metal	Advantages	Disadvantages
Recoverybycopying	V – volume of metal removed into chips; t_0– main time; v – cutting speed;b - cutting layer width	Steel	simplicity of design and adjustment of the cutting tool; possibility of processing wheels without dismantling some components	complex, expensive instrument; high roughness of the resulting surface; complex and highly qualified technology. equipment service
Surfacing	I -welding current; $V_н$ - surfacing rate; S - surfacingpitch; U_d - arc voltage	Steelgroups 1, 2, 3,4	reducing the wear of the band until its resource is completely exhausted, and the cost of servicing turning machines; elimination of idle time of wagons; salary savings boards	uneven heating of the surface; obtaining a deposited low-quality layer; preheating
Electricarcsurfacing	Current density 11-12 A/mm², preheating of the product to 200-400°C; tempering after surfacing up to 650...680 °C	Steelgroups 1, 2, 3,4	restoration of mainly small-sized parts; increase in wheel life	the formation of thermal failures due to uneven temperature fields; increased fragility of the deposited layer of the worn surface
Electroslagsurfacing	Welding current I, A. Electrode wire feed speed, m/h. Voltage U, V. Deposition rate v, m/h. Current density 0.2-300 A/mm²	Steelgroups 1, 2, 3,4	process stability over a wide range of current densities; high productivity; the ability to give the deposited metal the desired shape	thermal overheating of the thermally affected zone; special technological equipment, difficulties in fusing small parts of irregular shape

Consequently, all the restoration methods studied are applicable to this day, but it is necessary to conduct research to increase the durability of wheel sets using more economical, technological methods.

Thus, the service life of wheel pairs depends on a large number of factors, ranging from the design features of the profile of carriage wheels, the quality components of the wheel metal, operating conditions, and especially from the technology of restoration or repair. The service life of wheels restored by turning can be increased by minimizing turning operations and by reducing the thickness of the removed layer of metal of the rim and flange during each turning operation. It is necessary to optimize the parameters of the wheel processing process with a minimum amount of removal of the backing metal layer. It has been determined that the amount of wear on the tread surface of a solid carriage wheel of 1 mm occurs every 30-40 thousand kilometers. Accordingly, wear during an annual mileage will be ≈ 3 mm; and the metal consumption when turning the working part of the wheel rim, which goes into the chips, will be ≈36-43%. Consequently, to increase the service life of wheels, they strive to reduce the number of turnings and removal of the metal layer, which is achievable by the indicated methods.

Technological ways to increase the reliability and durability of wheel pairs involve restoring the tread surface and wheel flange using modern surfacing methods, including laser and plasma.

Research in the field of surfacing work to ensure durability and increase the overhaul life of carriage wheels has shown that there are significant shortcomings in the surfacing process.

Consequently, the studied technologies increase the service life of the tread surface and flanges of car wheels, but the surfacing process does not create sufficient protection of the hardened layer from the ingress of air and heating of the heat-affected zone. Therefore, the scientific and technical problem of creating a set of design and technological solutions to improve the quality parameters of the technological restoration process, ensuring an increase in the service life of the car wheel, requires a solution.

Thus, it has been established that the existing variety of recovery methods have similarities in general disadvantages. The main disadvantages that all methods have in common include: turning the rim reduces its thickness and shortens its service life; when exposed to heat, the elements of the base metal move into the penetration zone, reducing wear resistance; in most cases, preheating of the welded part is required, which can negatively affect its shape and size after cooling; a large allowance for finishing turning is required. A significant disadvantage of electric arc surfacing is that it increases the fragility of the metal of the restored surface due to the uneven distribution of temperature fields, which destroys it and leads to the formation of thermal cracks. Another feature of these technologies is that they provide for non-dismantling repair of wheels, which does not allow, under environmental conditions, to meet stringent requirements for the quality of welds and high physical and mechanical properties of the metal (RD VNIIZhT 27.05.01-2017).

When analyzing technologies for restoring car wheels, general technological parameters for restoration were determined. For example, under mechanical action, this is the volume of metal removed into the chips V, the main processing time t_0, the cutting speed of the tool v and the width of the cut layer b. When the wheel is exposed to heat, the general recovery parameters are welding current I; arc voltage U_d, current density J, deposition speed V_H and deposition step S. This means that in order to make a reasonable choice of the restoration method, it is necessary to focus on the main fundamental parameters that specify the directions and method of influence on the part. Justification of the qualitative parameters of the recovery method will be based on variations in general technological modes, changing their quantitative values.

Of the restoration technologies considered, the highest quality characteristics of the restored wheel are achieved by two methods: arc surfacing and laser surfacing with hardening. Both technologies require dismantling wheel sets and placing them in repair shops on special rotators. However, despite the advantages and quality characteristics of the methods, they have significant drawbacks, high overheating of the part and weakening of its phase structure, and the internal stresses present form a risk area for the formation of microcracks and wear.

3.2 Modeling the process of stress localization along the contact elements of the wheel in the Solidworks environment

Computer simulation modeling of the interaction between the wheel and the rail represents a complex multi-level contact system consisting of a large number of components that dynamically change over operating time [109-112].

The objective of the study was to assess the adequacy of the calculations and establish the limits of internal stresses in the wheel that arise on the tread surface of the wheel under the influence of cyclic dynamic loads. The problem was solved by modeling loading processes and the formation of stresses in the critical elements of the wheel. During the study, the SOLIDWORKS program was used using the Static II Pro package, etc.

It is known that the resource of any mating pair includes the stages of natural wear, steady wear (service life) and degrading wear (overhaul period). The result of any wear is the creation of a gap between the wheel, flange and rail, leading to increased dynamic forces and shock loads. This phenomenon is difficult to predict. Therefore, we assume that the wheel, axle, and frame of the body truck are technically sound and have passed a full technical examination.

The essence of computer modeling is to study fatigue stresses on contact surfaces. When modeling, the following conditions and priority factors were accepted: limits of values of the permissible deviation coefficient $X > Xmin$; repeating factors $Rol < Rop$ distortion of straightness at the critical point of the profile from the radius; undercut exception condition $Rol > 0$.

In the SOLIDWORKS program window, we construct the spatial coordinates of the application of the load along the cross-section of the contact area of the wheel and the rail and indicate the vector of application of the moments of force (Figure 34). We indicate the effective load in the transitional contact patch at the radial base of the ridge along the arc of the pitch diameter.

The area of formation of internal contact and bending stresses is visualized in the lower part of the wheel disk to a small extent. Spectral analysis of the color spectrum showed that the depth of stress distribution in the tread surface of the wheel and disk increases at the base of the shaft journal. The dynamics of stress distribution indicates that the shaft journal, wheel and disk operate under cyclic alternating loads that are formed in a short period of time t_σ. From the color diagram it can be seen that the section of the wheel under study is subject to loading $\sigma_F = 123,347 - 295,025$ MPa. The stresses present under dynamic loads are within the permissible criteria for strength conditions $\sigma_F \leq [\sigma_F]$.

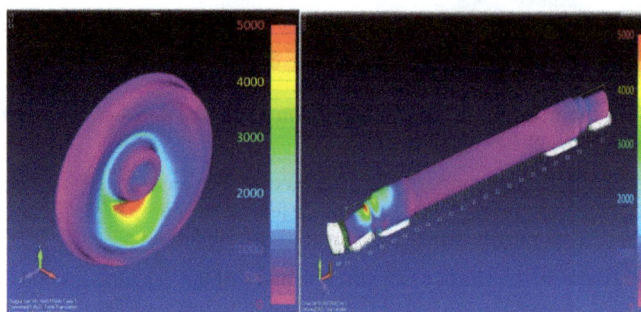

a)   *b)*

a) distribution of critical stresses in the wheel;
b) zones of stress formation along the axle journal

Figure 34. Principle of stress dislocation along the test part of the wheel

The movement of the car is largely realized through the transmission of torque to the traction wheel. The contact surface of the wheel first of all perceives the complex action of static and dynamic impact forces of resistance from its own masses (gross and laden) at different values of the load capacity coefficient and speed. It has been established that at the moment of acceleration and braking, especially on turns, the angular acceleration (ε_κ) and speed (ω) of the wheel reach critical maximum values. During period t_1 (acceleration time when the car is moving), the first dynamic contact of the wheel occurs, during which the internal stresses σ_F and σ_H intensively increase. During period t_1, the wheel and the transition zone of the flange contact the rail with a force of 48333 kN. Stresses in the upper layers of the wheel reach σ_F=123,347–295,025MPa. Stable movement with a constant speed of the car t_2 is characterized by the constant action of forces that are equally directed and small in value. We do not consider it in our study. The braking process $t_3$creates additional contact resistances that reduce the torque, but increase the moment of inertia. Critical voltage values are reached at the moment of rotation of the wheeled trolley along the rail radius t_4. Reaching critical stress values weakens the structure of the contact surface of the wheel, as a result of which the balance of the strength characteristics of the metal is disrupted. In this case, the action of a dynamic load due to alternating forces deforms the grains of the structure and breaks the molecular bonds, as a result of which particles peel off and the tread surface and wheel flange are deformed and worn out (Figure 35).

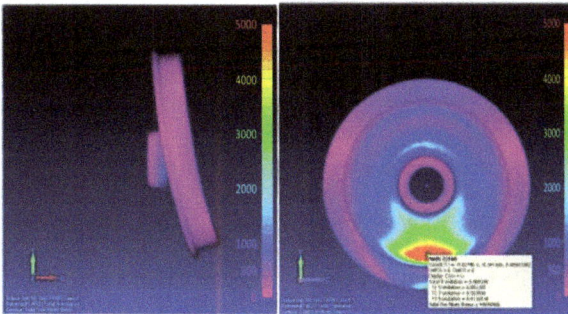

Figure 35. Influence of bending stresses on the mechanical properties of the wheel and its wear rate

The red color of the spectrum (Figure 35) shows that the effect of cyclic stresses is mainly observed on the axle and at the base of the wheel. Then the load mode and stress concentrators are redistributed along the contact surface of the wheel. At a critical moment, stresses cross the boundary state of the strength of the structure and lead to gouges or roll-ups and sliders. Instantaneous distribution of moments of forces and contact stresses σ_N over the contact area of the wheel, in a short time interval, increases the area of destruction of the contact surface of the wheel during braking.

The length of the wheel's travel during braking is accompanied by slippage. The described process is critical in nature and demonstrates boundary conditions under which the strength conditions of the structural phases of the metal are not met due to cyclically increasing contact and bending stresses over the wear area of the car wheel.

The processes under study explain the nature of destruction and compaction in the contact zone on the tread surface and the wheel flange. Structural hardening and surface rolling are observed in a constant region of stress concentration. The moment of the beginning of deformation of the contact layers coincides with the region of maximum accumulation of braking moments and at turns on curves of a railway car.

When choosing and developing an effective method for restoring car wheels using thermomechanical action, it is necessary to comply with the requirements for changing and modifying the mechanical properties of hardened wheels. The chemical-thermal, strength and mechanical characteristics of the restored surfaces of the wheel and flange must satisfy the optimal values: relative elongation δ of at least 10%, relative narrowing ψ of at least 16%; hardness $HB \leq$ 2430 MPa (248HB). It is possible to achieve them by substantiating the optimal laser exposure modes.

Car repair enterprises and research centers offer the manufacture of wheel structures from steels containing a minimum amount of alloying elements, but with increased resistance to the formation of fatigue and thermal cracks, structural strength, manufacturability, ductility, impact strength, increased heat resistance and wear resistance [113].

However, the research data in the field of changes in the profile of wheels, derived above in the work, do not take into account changes in the design geometry, shape and mechanical properties of the wheel during the operation of cars. Multidirectional vector forces and moments of inertia, at different speed modes and load capacity of cars, lead to accelerated degradation of the wheel tread surface. To understand the physical meaning of the formation of defects on the contact surface of the wheel and rail, it is necessary to solve the scientific problem of establishing the dependence of the influence of wear on the formation of wheel bending stresses. It is important to note that wheel wear (breaks, nicks, scratches) is probabilistic and very difficult to model. Therefore, let us consider a simplified design section of the wheel under study (Figure 36).

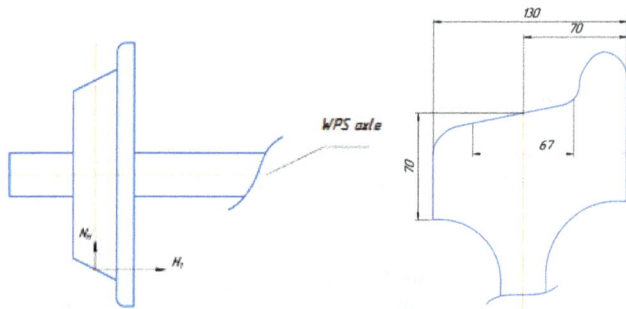

a) standard wagon wheel b) simplified diagram

Figure 36. Wheel design reduced to the design diagram

In Figure 36 (a) N_H is the vertical reaction of the rails (on the left wheel), H_1 is the lateral pressure. Having calculated the moments of forces and stresses acting in the wheel, we will construct N_z, σ_N, M_x diagrams.

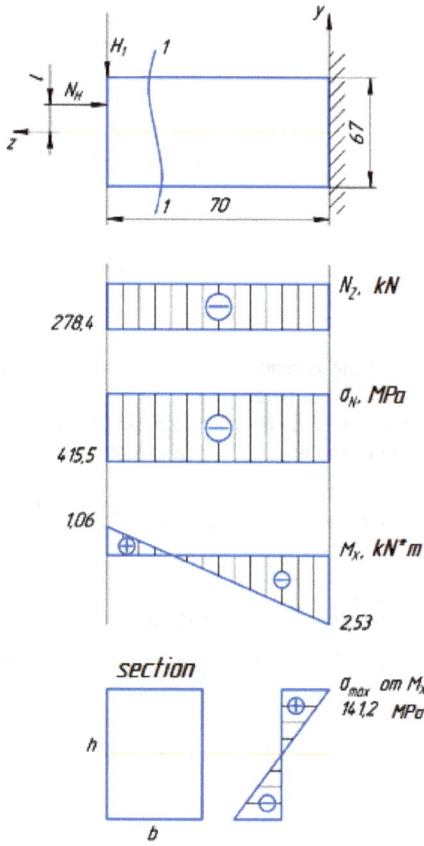

Figure 37. Diagrams of acting forces and stresses

We will calculate the values of the generated stresses in the contact zone of the wheel taking into account the action of the longitudinal N_H and transverse H_l forces (Figure 37):

$$\sigma = \sigma_N + \sigma_{Mx}, \tag{46}$$

whereσ_N – direction from longitudinal force (N_H),

σ_{Mx} – direction from shear force (H_l).

$$\sigma_N = \frac{N_z}{F},$$

where N_z – longitudinal force;

　F– cross-sectional area,

$\sigma_{Mx} = \frac{M_x}{I_x} \cdot y$- stress formed from a bending moment as a result of the action of longitudinal and transverse forces, M_x - bending moment, I_x-moment of inertia, y - coordinates along the section.

In an arbitrary section, we determine the longitudinal force:

$$-N_z - N_H = 0; \quad N_z = -N_H = -278,4 \text{ кN}.$$

It has been established that the contact between the wheel and the rail does not occur along a line, but there is a contact patch of $8 \div 13$ mm. For the study, we will take 10 mm, then the normal stresses will be σ_N = - $415.5 \cdot 10^6$ Pa. The (-) sign indicates that the stress forms a compression of the wheel material structure. Wheel flange contact area$F$ = 670 мм^2 = 0,00067 m^2.

The bending moment M_x is formed according to the principle (47)

$$M_X = +N_H \cdot l - H_1 \cdot z \tag{47}$$

at z=0; M_x = 1,058 кN·m;at$z$=0,07; M_x = -2,53 кN·m.

Let's calculate normal stresses during bending moments (Figure 38).

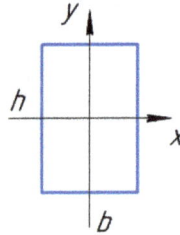

Figure 38. Conditional design cross-section of the wheel

$$\sigma_{Mx} = \frac{M_X}{I_X} \cdot y \tag{48}$$

where $I_X = \frac{bh^3}{12}$ – moment of inertia for a rectangular section about the x axis.

With h=67 mm and b=10 mm, Ix=2.51·10-7 m⁴; y=0.0335 m, then the desired σ_{Mx}=141,2 MPa. This means that at z=0 from the bending moment the total stress is σ= 556.7 MPa.

When the contact area during wear of the wheel tread surface in real operating conditions decreases by 2 mm ($h_1 = h - 2$ mm $= 67 - 2 = 65$ mm, F=670 mm²), the value of normal stressσ_N increases significantly $\sigma_N = -428,31$ MPa.

With a change in the worn contact area, the moment of inertia along the x axis also changes - I_x=2,29·10⁻⁷m⁴ and the stresses acting from bending moments increase σ_{Mx}=154,77MPa, the total stress in the contact patch - $\sigma = 583.08$ MPa (Figure 39).

$$I_X = \frac{0,01 \cdot 0,065^3}{12} = 2,29 \cdot 10^{-7} m^4.$$

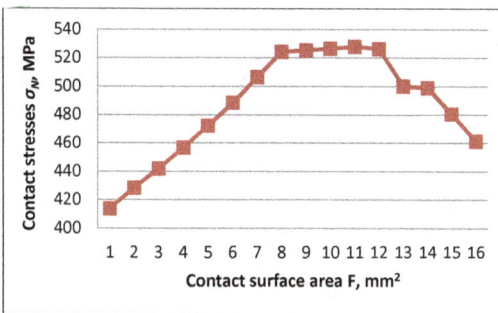

Figure 39. Dependence of the influence of changes in the contact area of wheel wear on the value of normal contact stresses

From the analysis of the graph it is clear that with an increase in the wear area of the wheel tread, the stresses that form fatigue wear increase. However, an increase in stress σ_N is observed up to σ_N =533 MPa, corresponding to the contact surface area F= 8 mm². This phenomenon describes the progressive wear of the contact surface of the wheel, which has a probabilistic nature [114]. With an increase in the wear area in the interval F = 8÷12 mm², stress stabilization is observed within the limits of σ_N = 527 MPa; this interval describes the process of an increased wear area, comparable to the surface of the entire rolling diameter. Consequently, the wear area is so large that it replaces the rolling surface and the stress decreases in magnitude. However, despite the decrease in σ_N, chipping of the rolling surface and deformation occurs, this is associated with wear of the hardened surface layer and, accordingly, a decrease in the physical and mechanical properties of the metal.

Dependencies have been established that describe the polynomial law of changes in normal stresses $\sigma_N = -1,502F^2 + 29,959F + 371,872$ and bending stresses $\sigma_{Mx} = 1,3749F^2 + 8,1023F + 132,6311$, respectively, on the wear of the contact surface of the wheel.

Consequently, the research results showed that when the contact area decreases by $2 \cdot 10^{-5}$ m^2 (20 mm), the stress value increases by 4,5%, which indicates the need to roll out and restore the wheelset.

Thus, the internal stresses σ_F and σ_H increase sharply in the initial stage of interaction of the contact pair of period t_1. The load will reach its maximum value during the final period of turning the car at radii t_4. The processes under study explain the nature of destruction and compaction in the contact zone on the tread surface and the wheel flange. Failures occur in the metal structure in the same zone of stress and deformation, corresponding to the maximum range of accumulation of moments and forces during braking and turning on curves of a railway car.

It has been proven that a decrease in the contact area during wear of the wheel tread surface by $h_i = 2$ mm at $\Delta h = 65$ mm, $F = 670$ mm^2, the value of the normal stress σ_N increases significantly $\sigma_N = -428,31$ MPa.

It has been established that when developing technology for restoring carriage wheels, the rim must maintain strength, high impact strength and wear resistance. The hub must be characterized by the required viscosity, the disc - elasticity. The chemical-thermal, strength and mechanical characteristics of the restored surfaces of the wheel and flange must satisfy the optimal values: relative elongation δ of at least 10%, relative narrowing ψ of at least 16%; hardness $HB \leq 2430$ MPa (248HB).

3.3 Development of a block diagram of an algorithm for substantiating criteria for the effectiveness of technology for restoring the profile of railway wheels

To increase the durability of wheel sets, conditions are necessary under which the railway system will operate uninterruptedly. This condition is to reduce the time of repair and restoration work to a minimum $t \rightarrow$ min, as well as to reduce downtime, because it worsens the efficiency criteria for the operation of wagons. The algorithm for increasing the time between repairs of a car wheel is presented in Figure 40.

In the course of research to increase the durability of car wheels, the following performance criteria were established: geometric, physical and mechanical, structural and operational (Table 6). These criteria show the load limits of wheelsets before and during operation of wheelsets.

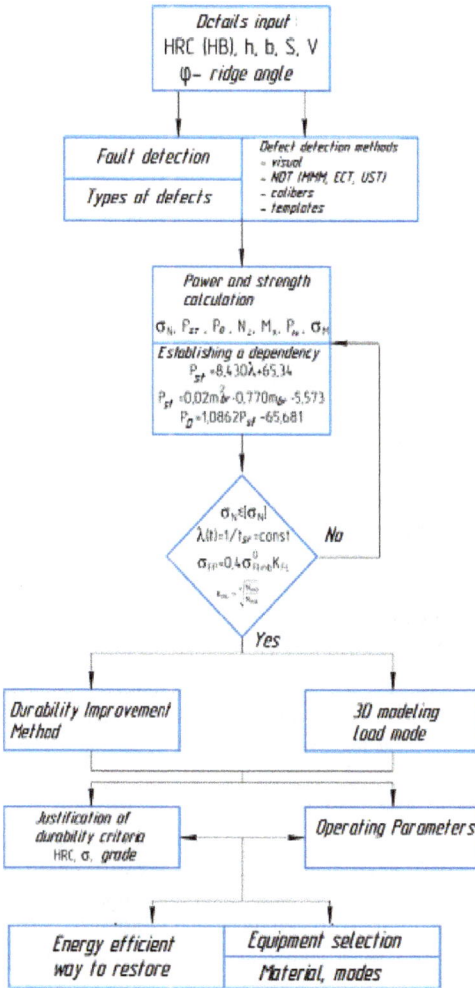

Figure 40. Block diagram of the algorithm for substantiating the criteria for the effectiveness of the technology for restoring carriage wheels

Table 6. Wheelsetefficiencycriteria

Geometric	Physicalandmechanical	Constructive	Exploitation
1. Wheel geometry: diameter – 950 mm, tread width ≈70 mm, rim width – 130 mm, flange height – 28 mm, flange width – 33 mm, roughness – *Ra* 40. 2. Shape: conical. 3. Type of bandage: bandageless (solid-rolled). 4. Rim diameter: 950 mm. 5. Comb angle:φ = 66°.	1. Hardness: from 280 HB to 360 HB 2. Material: steel 2 3. Permissible limits of fatigue stress: σ_F =123,347 – 295,025 MPa, relative extension – δ=8 %, relative narrowing – ψ=12%, temporary resistance – σ_u= 911-1107 MPa.	1. Type of axlebox: with plain bearings and with rolling bearings (roller) 2. The thickness of the disc should be optimal 3. Contact area of the wheel flange$F$= 670 mm^2 = 0,00067 m^2.	1. Cyclic stresses: static and dynamic loads – up to 25 ts. 2. Car speed$v$=15÷33 m/c; normal stresses will beσ_N=-415,5·10^6 Pa. 3. Operating conditions: braking, acceleration, steady motion, driving on turns and curves.

Thus, regardless of the high cost, versatility and possibility of modernizing the design of an industrial unit, its parts wear out during operation, and the main operational and technological parameters of restoration can vary widely and have different evaluation criteria. Therefore, the establishment of clear requirements for design characteristics that affect the physical and mechanical properties of wheels, as well as the establishment of basic technological parameters for restoring the wheel tread surface, require scientific justification and adjustment taking into account the listed algorithm requirements.

3.4 Experimental study of the quality of deposited material when varying the modes of laser restoration of a car wheel

Experimental studies were carried out using the method of planning a multifactorial experiment. To conduct an experimental study, select a parameter. Based on the selected criterion, we evaluate the carriage wheel. Next, we combine efficiency factors into a mathematical model.

As criteria for optimizing laser cladding, we define the following variables: y_1– hardness of the deposited layer, y_2 – strength, y_3 – adhesion (characterizing the strength and depth of adhesion). Before constructing the matrix of fused experiments, we select ten factors: x_1 – x_8, each of which varied at two levels: max and min. The list of factors taken into account is presented in Table 7. Then laser cladding was carried out. The studied surfacing parameters were chosen taking into account technological features and some properties of the material.

Based on the results of the experiment, we obtained a polynomial dependence that adequately characterizes the process of changing y from the selected criterion x_8. The dependence of y_1 (hardness) on x_8 (distance from the focal plane) looks like: $y_1 = -0,0119x^2 + 1,6283x + 2,8517$. Polynomial dependencies were similarly established$y_2$and y_3.

Table 7. Variable parameters of the mode during the laser cladding experiment

Modeparameters	Average values of selected parameters
x_1 – nickel, %	75
x_2 – silicon, %	4,5
x_3 – laser radiation speed, m/s	233
x_4 – laser radiation power, W	2500
x_5 – radiation focusing spot diameter, mm	2-2,2
x_6 – laser beam energy density, W/cm^2	$3\cdot10^5$
x_7 — wheel rotation speed, rpm	55
x_8 – distance from focal plane, mm	60-80

It has been established that the hardness and adhesion of the deposited layer, in addition to key modes (laser beam energy density, laser radiation power, wheel rotation speed, etc.), are also affected by such parameters as exposure time to the surface, distance from the focal plane and mechanical properties of the steel phase. It is substantiated that exceeding the distance from the focal plane by more than 80 mm leads to a decrease in the quality of surfacing (hardness and adhesion). This means that the established dependencies make it possible to optimize laser cladding modes when restoring the mechanical and geometric characteristics of car wheels.

Giving a worn wheel of rolling stock its original geometry during operation is a complex technological and organizational procedure in which a formed, worn wheel with fatigue stresses and defects is used as a blank. Classical methods, in addition to the well-known advantages, have significant disadvantages that negatively affect the properties of the metal structure of the wheel, the final shape and geometry after cooling, the formation of unacceptable cracks and the appearance of pores in the deposited layer, the fusion of the base and filler material. Carriage wheels are made from difficult-to-weld steel, which is not recommended for use in welding and surfacing.

Consequently, to ensure a restored surface of minimal deformation with high performance properties of the tread surface profile of a railway car wheel, on the other hand, it is necessary to develop an energy-efficient method for restoration by laser cladding.

The laser restoration method consists of local fusion of a layer onto the worn rolling surface of a carriage wheel by creating a minimum melting area of the base and filler material with high physical and mechanical properties (controlled by the chemical composition component of the deposited layer).

To select the main parameters of the mode and select the material for laser cladding, it is necessary to take into account the considered conditions of the restoration process. To do this, it is necessary to justify the optimal chemical-mechanical composition of the material used, as well as optimal parameters such as laser surfacing speed, radiation power, spot diameter, depth of the deposited layer, hardness after surfacing.

An experimental study for the selection of material, study of microstructure and microhardness for laser cladding of the running track of the rim of the wheel pairs of cars was carried out at the «Remplazma» LLP enterprise in Petropavlovsk.

It has been experimentally established that the structure, grain size, hardness and microhardness of the restored surface form such qualitative properties as wear resistance, stress concentration and adhesion strength of the coating to the base.

Thus, the optimal composition of a multicomponent powder composition for laser restoration can be practically selected by using various combinations of constituent elements, taking into account their properties and ability to prevent wear of the surface of the part during deposition. Methods for the practical preparation of such a composition may vary, depending on its purpose and use cases. The final indicators of the quality of surfacing using the specified composition are the durability of the subsequent operation of the wheel, as well as its overall efficiency.

In order to develop an energy-efficient method for restoration by laser cladding, it is necessary to substantiate the optimal coating application modes. For the study, samples of wheel steel measuring 15*30*15 mm with welded beads made of self-fluxing powder with the alloying element of nickel grade PG-SR2 or its replacement 15Cr17Ni12V3F (GOST 21448-75) were prepared. The granulation of the powder particles was chosen to be extra fine (EF) - from 40 to 160 microns.

To apply the laser heat-strengthened coating, we used a MUL-1 laser welding installation from Laticom - Laser Technologies and Components LLC, manufactured in Moscow, Russia, containing a pulsed solid-state laser with a radiation power of 3 kW.

Laser surfacing was carried out at surfacing speeds v=10, 15, 20 mm/s, at a surfacing distance l=10.12 and 15 mm and radiation power P=1000-3000 W when scanning a laser beam in one pass. Next, the deposited samples were cut perpendicular to the direction of deposition, and microsections of the deposited layer were prepared to determine the geometric parameters of the layer: width b and height (depth) h of the beads. Measurements of these parameters were carried out using a Micro P200 optical microscope.

Processing of experimental data was carried out according to the method of a full factorial experiment, using electronic tables.

Microhardness studies were carried out using a PMT-3 microhardness tester. In Figure 41, a shows a microsection of one deposited track, and the overlap zone of the deposited tracks is shown in Figure 41, b.

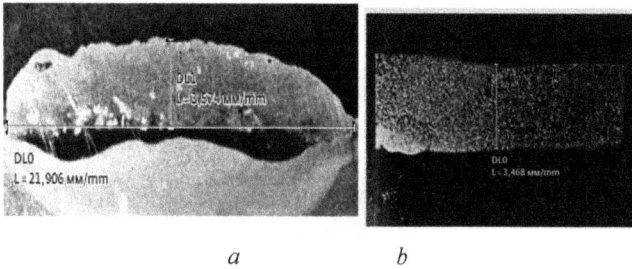

a   b

a - single deposition path in one pass, b - zone of overlap of deposited layers (transition)

Figure 41. Study of deposited material on samples

The deposited beads have a continuous dense composition with clearly defined structural components. The deposited coating consists of a zone of deposited powder (fusion zone), a thermally affected zone, and a base metal zone. The formation of the first zone occurs at high temperature and high cooling rate of the weld bead.

Figure 42 shows the change in microhardness at different deposition rates depending on the thickness of the deposited bead.

Analyzing the microhardness study graphs, a sharp jump in the growth of microhardness is observed during the transition from the base metal to the weld bead, indicating a small size of the transition zone. With a further coating thickness of 0.8-1.5 mm, the microhardness is almost at the same level of 5550 MPa, which indicates the homogeneity of the material and the uniformity of temperature distribution along the 10 mm cut of the roller. At a distance of 10-15 mm and a deposition speed of 15 mm/s, we observe high microhardness, and low microhardness is observed when the deposition speed is reduced to 10 mm/s.c.

Optimal microhardness values of 5000 MPa were achieved thanks to an optimal fusion transition zone of 0.8 – 1.45 mm and a laser beam diameter of up to 1.3. High microhardness was achieved with an optimally justified distance of 10-15 mm and a deposition speed of 15 mm/s. A further increase in the transition zone to 1.45 mm does not significantly change the microhardness, and with an increase in thickness to 1.6 mm, the microhardness sharply decreases to 3974 MPa.

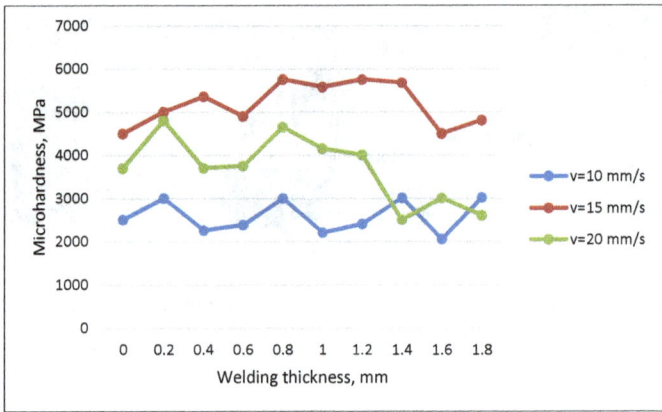

a)

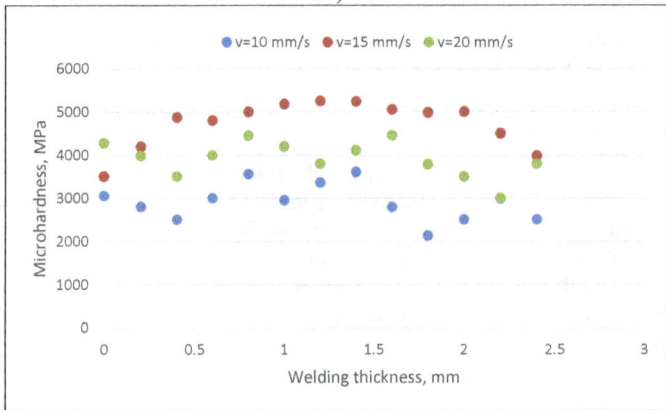

b)

Figure 42. Microhardness measurement along the thickness of the deposited layer at a distance of 10 (a) and 15 (b) mm at different deposition rates

A regression equation has been established for the studied dependences of microhardness on the thickness of the deposited layer at a surfacing speed of 15 mm/s; it is described as $y=-977,58x^2+2499,7x+3689,5$ with a correlation coefficient $R^2= 0.9314$, at a speed of 10 mm/s the regression equation for this dependence was determined $y=-1,2083x^2+44,2312x+2592,1$ cR^2=0,98 with R^2=0.98.

Thus, these studies indicate the need for a preliminary study of the surfacing process with wear-resistant filler material to obtain the required high-quality coating. The established patterns make it possible to select materials and select laser surfacing modes to obtain optimal coating properties from the PG-SR2 alloy (15Cr17Ni12V3F). When choosing the optimal laser cladding modes, the values of the radiation power, the speed and focal length of the cladding, and the diameter of the laser beam were changed. At a deposition speed of 15 mm/s, the microhardness varied within a wide range of 5000–6000 MPa in the deposited layer, and this speed was determined to be optimal. By changing the power density of the laser beam energy, it was determined that with an increase, the microhardness decreases, thermal defects arise, with a decrease, the microhardness of the deposited layers also decreases and incomplete fusion of the powder material occurs. Also, an increase in microhardness occurs with a decrease in the deposition distance.

Consequently, as a result of the experimental study, the parameters and modes of laser cladding were established:

- As a filler material we use PG-SR2 powder (15Cr17Ni12V3F) with an alloying element - nickel of the *Ni-Cr-B-Si* system (Table 3.4), it gives high hardness *HRC 40*, corrosion resistance and heat resistance, and has a low melting point (960...1000 °C), which helps to reduce the thermal impact on the base metal and reduce the level of residual deformations and stresses at the base of the wheel. The selected powder at high processing temperatures forms a glassy slag coating that protects the deposited coating from interaction with oxygen and nitrogen in the air, eliminating the need for special protection from negative environmental influences.

Table 8. Chemical composition of nickel-based powders for laser technology

Alloygrade	C, %	Cr, %	Si, %	B, %	Fe, %	HRC
PG-SR2	0,2...0,5	12...15	2...3	1,5...2,1	lessthan 5	35
PG-SR3	0,4...0,7	13,5...16,5	2,5...3,5	2...2,8	lessthan 5	45
PG-SR4	0,6...1	15...18	3...4,5	2,8...3,8	lessthan 5	55

- radiationpower 2000...3000 W;

- deposition speed 10-15 mm/s;

- radiation focusing spot diameter 1.5...2.5 mm;

- diameter of the processing laser spot (cladding distance) – 10-20 mm;

- laser beam energy density - $3 \cdot 105$ W/cm^2;

- distance from the focal plane – 20-100 mm;

- mass flow rate of surfacing powder 0.25 g/s.

In the deposited layer, microcracks may form in the overlap area of the rollers, which can be avoided by preheating and slow cooling of the deposited wheel tread surface. The hardness of the deposited layer should be HRC 30...40. It is also possible to apply a matte coating (MCS-510, SG-504, FS-1M) [115] with high absorption capacity, non-toxic, non-flammable, easy to apply, inexpensive.

The recommended laser restoration method ensures more efficient fusion of base metal particles with the tread surface being restored at the atomic level and the necessary wear resistance, and by adjusting the operating characteristics of the restoration process, the physical and mechanical properties of the wheel can be significantly improved. Laser coatings have higher physical and mechanical properties compared to coatings applied by classical surfacing methods.

Deformed, worn-out wheelsets based on quality indicators that are not subject to rejection are necessarily either repaired or restored. The classic technological process is carried out without changing elements and consists of standard stages according to a typical method: input control with depressing from the axis, restoration of the treadmill profile and output control[116].

The process of restoring the wheelsets of railway cars in our study consists of the following:

- loading - lifting the car using a jack, moving and securing the wheel pair onto a retractable ramp, installing them on a mobile platform, moving them to the welding booth;

- cleaning from grease, dirt, rust, oil, paint residues of wheel pair elements, while cleaning is carried out either with detergents or scrapers, metal brushes, rags, sandpapers;

- flaw detection for wear, changes in design geometry, the presence of cracks, sliders, rolled products, gouges, pointed ridges and other defects (the type of defect, identification method, division of defects into 2 categories (acceptable and unacceptable), measuring instrument, recommended repair methods and permissible limit values after repair restoration);

- preparation for turning (bringing the wheel lathe to the wheel being processed, setting the program for turning the tread surface);

- turning the wheel tread (setting the rotation speed in the control unit of the rotating mechanism and cutting modes on the machine);

- preparation for laser cladding (approach and installation of the laser head to the treated wheel tread surface);

- establishment of modes for laser deposition (set deposition modes for coating on the MUL-1 installation);

- powder deposition - for shallow faults of wheels (base material - PG-CP2 powder (15Cr17Ni12V3F) on the restored tread and flange surface, MUL-1 laser deposition installation, PG-CP2 powder particle size 1064 microns, spraying distance 100 mm, laser size weld is 0.2-2.5 mm, pulse repetition frequency 120 Hz) and wire fusion (LP-Ni, diameter 0.2-0.8 mm) - for deeper defects in car wheels;

- defect detection (NDT check for the presence of microcracks on the surface of the running track of a carriage wheel);

- turning operation (bringing the wheel to its original geometry and roughness);

- defect detection (checking the NDT for cracks, stress in the wheel tread, design profile of the contact surface, angle of flange transition);

- unloading (installation of the restored wheelset under the rolling stock car).

Thus, to ensure the durability of the contact surface of the wheel, it is necessary to create an optimal microhardness of the deposited layer, which ensures high wear resistance of the wheel. Optimal microhardness values of 5000 MPa were achieved thanks to the optimal transition fusion zone of 0.2 – 0.25 mm. High microhardness was achieved with an optimally justified distance of 60-80 mm and a deposition speed of 15 mm/s. A further increase in the transition zone to 0.35 mm does not significantly change the microhardness, and with an increase in depth to 0.5 mm, the microhardness sharply decreases.

Having established the optimal microstructure by metallographic studies, it was possible to substantiate the optimal modes for restoring the tread surface of the wheel and flange: radiation power 2000-3000 W; surfacing speed 10-15 mm/s; radiation focusing spot diameter 1.5-2.5 mm; diameter of the processing laser spot (cladding distance) – 10-20 mm; laser beam energy density - $3 \cdot 105$ W/cm2; distance from the focal plane – 50-200 mm; mass flow rate of surfacing powder is 0.25 g/s.

The use of laser technology for the restoration of car wheels using the developed method increases the time between repairs of the wheel, its efficiency increases when using production equipment. The creation of a mobile complex with powerful technological equipment for laser restoration will reduce car downtime and the cost of restoration work while simultaneously improving the quality of the restored wheel, which will help increase its competitiveness. Experimental studies to improve the wear resistance of wheel steel metal have shown that the hardened layer by laser cladding increases the hardness of wheel steel, thereby reducing wear by 7%.

Conclusion

The technology of mechanical turning of the deposited coating of the rolling surface and the flange of a wagon wheel is not regulated by instructions; wagon enterprises use their own production know-how and devices, but this does not guarantee their high energy efficiency; therefore, they form a significant share of the costs of the restoration and repair process.

The highest quality characteristics of a restored wheel are achieved by arc surfacing, plasma and laser surfacing with hardening.

Studies of the wear resistance of deposited layers have shown that when laser surfacing of products operating under conditions of intense wear, it is necessary to use materials so that the hardness of the deposited layer is HRC 35-40. The study found that it is necessary to use filler materials with high wear resistance. Nickel-based materials are defined as such filler materials. Therefore, in modern rolling stock repair practice, scientific research is being conducted to develop new energy-efficient methods for restoring the tread surfaces and flanges of carriage wheels using wear-resistant filler materials.

The optimal modes for restoring the tread surface of the wheel and flange have been established and justified: radiation power 2000-3000 W; surfacing speed 10-15 mm/s; radiation focusing spot diameter 1.5-2.5 mm; diameter of the processing laser spot (cladding distance) – 10-20 mm; laser beam energy density - $3 \cdot 105$ W/cm^2; distance from the focal plane – 20-100 mm; mass flow rate of surfacing powder 0.25 g/s.

Consequently, the scientific problem of developing a systematic approach to studying the cause and establishing the dependence of changes in wear on the wheel contact surface on dynamic loads and the intensity of displacement of the contact spot from the axis of the car's trajectory on curved sections has been solved.

Laser cladding for restoring and increasing the durability of railway wheels   Materials Research Forum LLC
Materials Research Foundations **157** (2024)   https://doi.org/10.21741/9781644902912

Chapter 4. Development of a Mobile Complex and a Method for Laser Restoration of Wheel Sets of Railway Cars with Increased Durability

4.1 Theoretical analysis of existing mobile surfacing installations

In modern times, the use of rolling stock not only for passenger transportation, but also for freight, leads to wear and frequent failures of railway car components precisely in the way of movement, i.e. where prompt repair of wagon wheelsets is required. These failures occur in field, off-site, off-site, remote from the MP places. For this purpose, repair land and railway complexes have found application. The efficiency, versatility and use of complexes for repair work with the presence on its platform of a base of machine tools for turning is correlated with their significant quality – mobility or non-stationary arrival. These complexes expand the possibilities of restoration work on turning and mobile repair, as well as effective management of repair and restoration cycles. The next advantage is the access of the supply of the turning or welding-surfacing equipment to the wheelset itself directly onto the rail track on any part of the movement and independently with the uncoupling of the car, or without uncoupling, only with the mandatory presence of power on the platform [117].

These design features of the mobile complex allow for immediate repair work in case of failure of any components without rolling back the entire rolling stock or a single car to repair shops.

In order to prevent critical moments of rolling stock operation and timely correction of defects in wagon wheels, our research is aimed at finding equipment and developing a complex for restoring railway wheels, which is applicable in case of an unplanned breakdown at a distance from repair points. There are many identical complexes, but well-known repair complexes have flaws.

Studies have established that mobile repair complexes that ensure prompt arrival at the repair site and high quality restoration of the design geometry of the ridge and rim practically do not exist, and the available analogues are extremely limited in functionality.

A mobile welding complex is known (Figure 43) based on the chassis of a car with a pneumatic motor. The design of the welding complex consists of a driver's cab with machine tools, a working module including a welding machine, an autonomous diesel generator, a pumping station, a control cabinet and a manipulator for moving the welding machine, as well as a control panel on its base platform (RU 131733 U1, 2013).

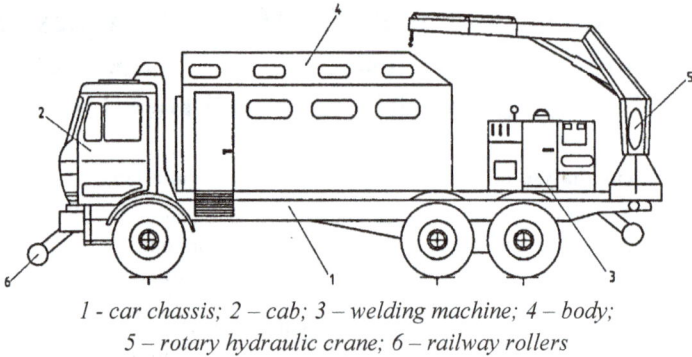

1 - car chassis; 2 – cab; 3 – welding machine; 4 – body;
5 – rotary hydraulic crane; 6 – railway rollers

Figure 43. Mobile welding complex

The disadvantage of such a complex is the limited functionality, the lack of equipment for restoring wheelsets, the absence of an independent repair zone, applicability only for welding railway rails, the absence of boring modules, the inability to technologically improve the physical and mechanical properties of the contact surface of the rolling wheel, the absence of a mechanism for turning the wheelset, rotation mechanism and rigging platforms.

Figure 44 shows a self-propelled repair complex of the RK-1M brand. The investigated repair complex is designed to replace spare wheelsets by removing them from the base of the complex, installing a lift of wagons and serviceable WPS on the rail track, then loading defective faulty worn-out WPS and lift on the platform of the complex. The design features of the complex consist in the presence of a main platform containing a means of movement in the form of a frame with wheels and a drive, the presence of a lift of wagons with supports on the ground behind the dimensions of the sleepers of the rail track. A special feature of the platform is a rotary manipulator crane with an arrow for loading and unloading the WPS. There is also additional equipment in the form of a set of devices, an energy supply system and a backup spare WPS (RU 102925 U1, 2010).

Figure 44. Repair railway complex RK-1M

The disadvantage of this complex is structural, technological and functional limitations: the absence of a retractable loading platform, a mobile welding and surfacing installation, the absence of a repair cabin with climate control over the welding post and the absence of a lifting stack for the restored set of wheels, not full use of the potential of repair and restoration work and the inability to restore the surface of the wheels.

An improved model of repair complexes for the restoration of wheelsets of rolling stock is an installation complex with a mechanism for hanging out a wheelset in the form of a self-propelled electric jack mounted on a trolley placed on side railway tracks laid inside the main track, a separate trolley with an arc ignition device, a drive for feeding welding wire and flux into the arc zone and a device for turning fused wheels [118].

The disadvantage of the presented complex is the need to install additional railway tracks, which is not always possible when working on the track away from the repair depot, the absence of a welding booth with climate control over the welding post, the inability to automatically control the process of restoring the design geometry and physical and mechanical properties of the skating surface with subsequent removal of internal stresses, which reduces the quality of the restored skating surface when working outdoors.

Consequently, these models have significant shortcomings in the design technology and selection of the necessary stationary equipment and equipment. The objective of our research is the scientific and technical solution of the identified problems with the creation of a single mobile complex that will solve the technological, operational, functional, repair and restoration disadvantages of previous inventions.

In modern times, there is a need for rapid turning of wheel sets of wagons in various conditions, therefore mobile wheel-turning machines that are easy to deliver to the repair site are actively spreading, and there is no need to transport a wheel pair to a specialized depot.

Thus, when repairing wheel sets of rolling stock, there are three ways of restoration: by uncoupling repair of wagons in the train at repair sites and outside them; by uncoupling repair of wagons; by uncoupling repair with rolling along the railway track of the mobile complex. Uncoupling and uncoupling repairs are carried out with the help of the self-propelled repair complex RK-1M, lifting the car and rolling out the chassis to the technological platform, carrying out welding work or replacing wheel pairs. The uncoupling of the car from the train is carried out to drive it to the nearest MP, and the train itself - to a railway dead end in order to continue moving after the repair. The latter way is the most relevant, reducing the downtime of rolling stock and shortening the time of either restoration or replacement of nodes.

4.2   Development of a mobile complex for laser restoration of wheel sets of wagons

Despite the insignificant wear area of the wheel contact spot, the moments of inertia and contact stresses increase significantly, which leads to thermal destruction and deformation of the wheel metal. This task will be solved by the development of a mobile repair complex that restores the original design geometry of the worn ridge (Figure 45) and the treadmill of the wheel and modifies the high physical and mechanical properties of the rolling surface with optimal stresses in the structure of the base metal of the wheel, as well as increases the efficiency and quality of restoration of worn wheels of railway cars (Figure 46).

Figure 45. The worn-out crest of the wheelset, restored by the developed mobile complex

Figure 46. Experimental work with simulation of field conditions on the premises of the «Remplasma» LLP enterprise

The task provided by the designed model is to increase the operational efficiency of the mobile repair complex by increasing productivity, reducing the complexity of work and expanding its functionality, as well as improving the quality of restoration of the rolling surface and the crest of wagon wheels.

The result was achieved by the development of a multifunctional mobile platform containing a retractable fork platform, winch-traction modules, an autonomous hydraulic drive system, a full-turn rotator on which lifting cylinders with high lifting capacity with roller cups are installed to reduce friction when the axis rotates, an upper divided platform – a storage device, a lifting stacker for manipulating repaired wheelsets, a multi-axis welding post, a welding cabin, which houses universal surfacing and milling-boring equipment, thermal shut-off heaters that support the climatic conditions necessary for optimal surface modification during surfacing. Unification of the equipment complex directly on the platform eliminates the need to install additional tracks and reduces the time of operations.

It is possible to ensure the achievement of a constructive technological result by the fact that a full-rotation rotator is installed on the mobile platform in its rear part, which allows you to automatically adjust the coordinate location of the repaired wheelset relative to the surfacing nozzle as efficiently as possible and select the optimal angle and distance of surfacing, and four hydraulic lifts are fixed on the diametrical side, on the tops of which the locking rollers ensure the rotation of the axis suspended state; a milling-wheel cutter is installed nearby on the turntable in the pre-end space, which ensures the cutting of the rolling surface immediately after surfacing, removing surface stresses; the full-turn platform is powered by an installed hydraulic motor and a planetary gearbox, and a drive gear is fixed on the output shaft, transmitting additional force to the crown of the turntable, thereby changing the position the wheeled trolley relative to the recovered nozzle; in the rear part of the platform there is a recovery cabin with through entrances overlapping with roller blinds; in the upper parts of the cabin there are thermal shut-off units that control the microclimate near the welding bath, creating a temperature regime and humidity level; mobility and laying of the restored wheelsets is provided by a lifting stacker installed in the middle part of the platform.

Figure 47 shows the general scheme of the mobile complex, which restores the design geometry of the worn ridge and treadmill of the wheel and modifies the high physical and mechanical properties of the riding surface with optimal stresses in the structure of the wheel base.

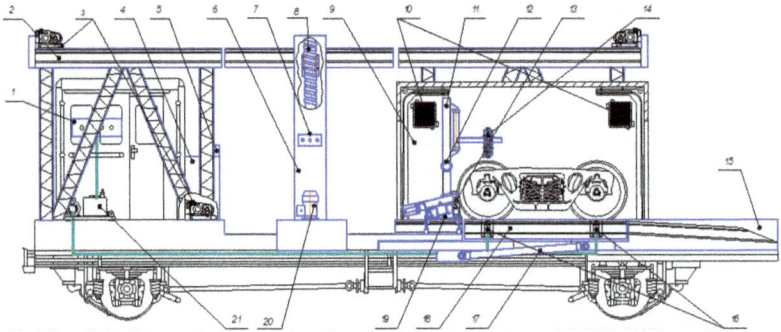

1 - hydraulic station control panel; 2 - upper tier;
3 - electric winch; 4 - welding transformer; 5 - control unit of the welding machine; 6 -
electric stacker; 7 - control panel of the stacker; 8 - screw rod of the lift;
9 - welding booth; 10 - heat shut-off heater; 11 - crane beam; 12 - handle;
13 - rotary mechanism; 14 – laser head; 15 - retractable ramp; 16 - hydraulic jack;
17 - hydraulic rod of the hydraulic cylinder of the retractable platform;
18 - full-turn circle; 19 - wheel milling and boring device; 20 - electric drive;
21 - hydraulic station

Figure 47. Mobile complex for the restoration of wheel sets of railway cars

The mobile repair complex for wheel restoration contains a hydraulic station 21 with a control panel 1, driving hydraulic jacks 16, a hydraulic rod 17 of a hydraulic cylinder of a retractable platform and a full-turn circle 18. The complex has an upper tier 2, to which an additional or restored wheeled trolley can be moved using an electric stacker 6. Electric winches 3 are installed on the upper tier, which pull the trolley from the stacker onto the platform using rail guides. Electric stacker, equipped with an electric drive 20, a screw rod 8 and a control panel 7.

There is a welding booth 9 on the platform of the mobile complex. The functionality of the welding booth allows for restoration work using a crane beam 11, with multi-coordinate surfacing equipment fixed to it, consisting of a laser head 14 with a device for feeding wire or powder and gas. In addition to the equipment for surfacing, there are also welding devices - a welding wire cassette and a hopper with a flux for welding, a control unit 5 and a welding transformer 4. The crane beam has several degrees of freedom and can be adjusted in height, using the handle 12, and at a certain angle. Also, the welding cabin is equipped with a mechanism 13 that rotates the wheelset, a milling-boring device 19 and heat shut-off heaters 10. In addition, the complex has a retractable ramp 15, which is necessary to lift the wheeled trolley onto the platform.

The mobile repair complex works as follows. The lifting of the car is carried out using four jacks (not shown in the figure). We select single-column portable jacks with the possibility of fixing them on different sides opposite the pin beams of the car.

Then, lifting the car, the wheeled trolley rolls out from under the car and moves, with the help of an electric winch 3, along the rail guides of the retractable ramp 15 to the full-turn circle 18 (Figure 4.6).

Ensuring loading of the wheelset onto the platform requires the introduction of a winch mechanism. To design an electric winch, it is necessary to calculate its power and traction characteristics.

The maximum tension of the rope is determined from the equality:

$$F_{max} = \frac{G}{Z_{к.b.} u_p \eta_p \eta_{n.bl}},$$

where G – wheelset weight;

     $Z_{к.b.}$ – the number of rope branches wound on the drum;

     u_p – the gear ratio of the polispast;

     η_p – The efficiency of the polispast;

     $\eta_{n.bl}$ – Efficiency of guide blocks.

The type of rope is set from the condition $F_{max} z_s \leq F_{br}$, where F_{br} is the breaking force of the rope; z_s is the safety factor, we select for the M7 operating mode

The diameters of the blocks D_{bl} and the diameter of the drum D are compared depending on the diameter of the rope d_k according to the conditions $D \geq h_1 d_k$; $D_{bl} \geq h_2 d_k$; $D_{ur.bl} \geq h_3 d_k$, where h_1, h_2, h_3 are the coefficients for choosing the diameters of the drum, block and equalizing block for the M7 operating mode

The length of the drum l_b, the thickness of the drum wall δ and the diameter of the axis of the smooth drum will be set by equalities:

$$l_b = \frac{L_k d_k}{\pi z (D + z d_k)},$$

where L_k – total rope length;

     z – number of winding layers.

The wall thickness is equal to $\delta = 0{,}01D + 3$ mm, and it must be checked for strength and compression according to the allowable compressive stress $[\sigma]_{cj}$, given in Table 9.

Table 9. The given values of the compression tolerance in the design of the winch mechanism

Brand of winch drum material	$[\sigma]_{cj}$ for the mode of operation with a wheelset - M7
St3sp	130
20	140
15XSND	175

The calculation of the bending moment M_i and the allowable stress $[\sigma]_i$ and by the following inequality $d \geq \sqrt[3]{M_i/[\sigma]_i}$ will determine the optimal diameter of the drum axis. The choice of the winch mechanism drive is implemented depending on the rated power of the engine $P_{dv} = (0,7...0,8)Gv/\eta_{pr}$, where η_{pr} is the preliminary value of the efficiency of the winch and the gear mechanism, and the gear ratio of the gearbox should not differ from the desired gear ratio under the condition $(u_{r.is} - u_p/u_{r.is})100\% \leq 15\%$.

The characteristics of the effective braking of the winch mechanism are considered in the ratio of the nominal from the calculated braking moments according to the condition - $T_{m.n} \geq T_{m.p}$, where $T_{\text{т.р.}} = K_{\text{т}} \dfrac{DG\eta}{2u_{\text{Mex}}}$ the full gear ratio of the winch mechanism u_{Mex}.

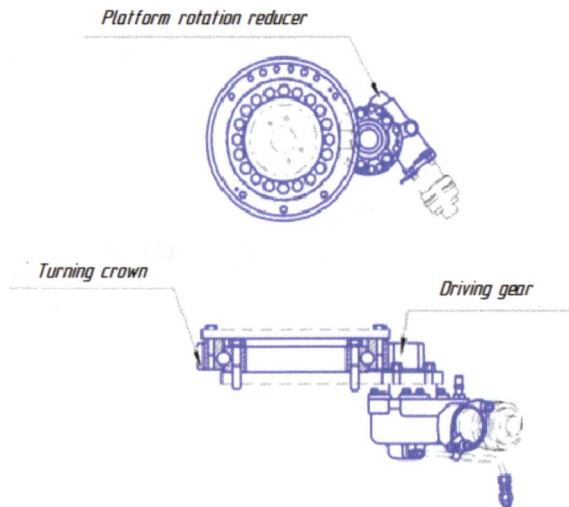

Platform rotation reducer

Turning crown

Driving gear

Figure 48. Image of a full-turn circle for installing a wheelset

The mobile complex is a complex power system. The operation of the drive for the dynamics of the main structural elements requires effort. The proposed full-turn platform rotates due to the installed hydraulic motor and planetary gearbox (Figure 48).

The further process is realized by rotating the ramp. The ramp, driven by hydraulic rods 17, moves along the rollers (not shown in the drawing). Hydraulic jacks 16 are installed in the turntable, which raise the processed wheelset to the desired height (Figure 48). The multi-coordinate laser head 14 is installed in the working position using a mobile crane beam 11. The surfacing modes are set on the control unit 5, the power comes from the welding transformer 4. The wheelset is driven by a rotary mechanism 13 (Figure 49), on which the required rotation speed is set.

Figure 49. Rotary mechanism

The process of laser surfacing occurs according to the scheme shown in Figure 50. The supply and transportation of planting material from the feeder to the distributor and then into the coaxial nozzle is realized under gas pressure. In the process of moving the laser head, we weld a layer of filler material onto the worn surface.

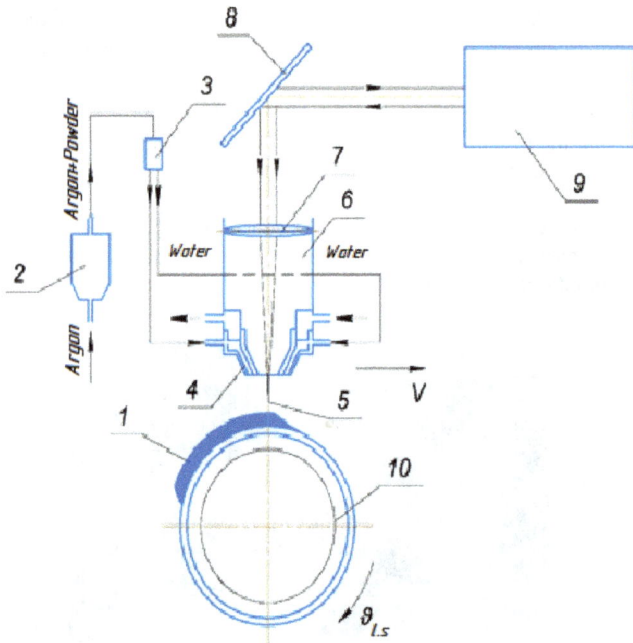

1 – surfaced layer; 2 - feeder; 3 – distributor of gas–powder mixture; 4 – coaxial nozzle; 5 – laser beam; 6 - surfacing head–lens; 7 – lens; 8 – rotary mirror; 9 - laser installation; 10 – wheel

Figure 50. Laser surfacing scheme

To ensure the optimization of the quality indicators of laser surfacing, the process takes place in a consumable laser beam (Figure 51). The diameter of the spot d_P is selected depending on the change in the defocusing distance ΔF. Using the maximum pumping energy depending on the pumping voltage of the laser installation, thereby increasing the uniformity of the energy distribution over the spot. To select the radiation power, we change the spot diameter and pulse power. The laser surfacing zone is expressed by the depth h of the penetration and the width b of the spot. It is established that the width of the LIZ is incomparable with the diameter of the spot.

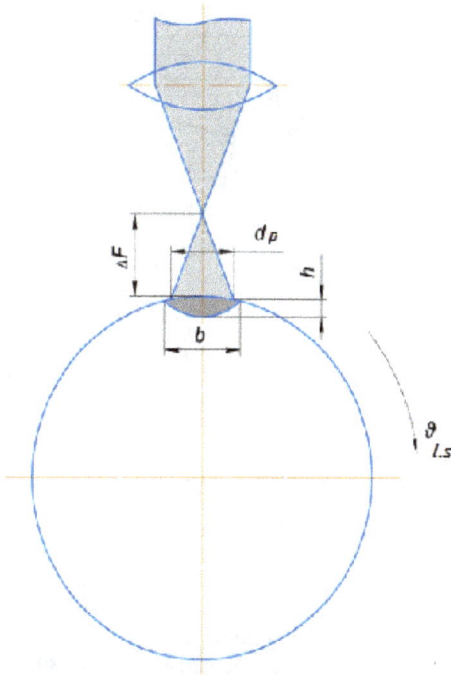

Figure 51. Dimensional scheme of laser surfacing

Next, the boring of the rolling surface to the specified dimensions is carried out by a wheel milling and boring device 19 (Figure 52).

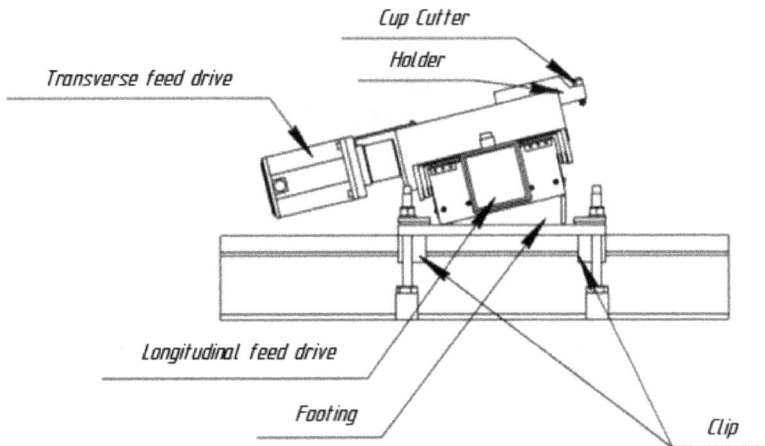

Figure 52. Mobile wheel boring machine

Surfacing and boring takes place in a welding booth, in which heaters maintain optimal temperature and humidity above the welding bath, installed above the entrances. To restore the opposite part of the wheelset, the cart is rotated 180° using a full-turn circle 18.

Thus, the use of the claimed mobile recovery complex provides the possibility of prompt arrival at the work site, high quality restoration of the design geometry of the ridge, treadmill and high physical and mechanical properties of the modified rolling surface of the wheelset, the ability to automatically adjust the optimal operating parameters of recovery, which leads to an increase in the efficiency and quality of restoration of worn wheels of railway cars without installation additional railway tracks

4.3   Development of laser technology for restoring the rolling surface and the crest of wheel pairs

The development of high-quality and effective laser surfacing technology is a complex and multifactorial task associated with a large amount of experimental work. To obtain optimal recovery modes, it is necessary to determine experimentally the qualitative indicators of laser surfacing: the required depth and width of the hardened layer, the absence of cracks and metal melting, while the deposition should be carried out in 1 pass, the deposition zone should be closed and continuous.

Studies of laser surfacing quality indicators were carried out at the Department of Transport and Mechanical Engineering of M. Kozybayev SKU in the laboratory «Non-destructive Testing and evaluation of physical and mechanical properties of parts», as well as during a scientific internship in SibADI and OMSTU, Omsk, Russia, using the latest energy-efficient equipment.

The samples were studied at the MUL-1 laser surfacing unit, which is a pulsed solid-state laser operating at a wavelength equal to 1064 nm. Figure 53 shows the appearance of the laser surfacing installation. Structurally, the laser installation is made in the form of two devices – a laser emitter with a mechanization system, a microscope and a welding head (optical module) and a housing including a cooling system and a power source (power module).

Figure 53. Appearance of the MUL-1 laser surfacing unit

The MUL-1 laser processing unit has high power and good performance. The modular design feature of the installation makes it compact and mobile, as well as having extensive technical characteristics and parameters (Table 10).

Table 10. Technical parameters and characteristics of the MUL-1 laser installation

Technicalspecifications	Numericvalue	Unit of measurement
Overalldimensions	540*1100*1250	mm*mm*mm
Power supply and cooling system	660*790*360	mm*mm*mm
Opticalunit	700*500*360	mm*mm*mm
Laser type - based on Nd:YAG with tube pumping, wavelength	1064	nm
Maximumradiationpower	100	w
Averageradiationpower	50	w
Pulseduration	0,4...20	ms
Pulse repetition rate (the duration of the pulse series is unlimited)	0,2...20	Hz
Maximumpulseenergy	80	J
Maximum pulse power (peak power)	10	kW
Total weight of the installation	70	ru
Power supply	220; 50	V; Hz
Maximumpowerconsumption	nomorethan 2500	W

The main parameters and technical characteristics of a laser built on the basis of a crystal of yttrium-aluminum garnet with neodymium (Nd:YAG) tube pumping are presented in Table 11.

Table 11. Main experimental parameters of the laser beam of the pulsed solid-state laser MUL-1

Parameter	Designation	Numericvalue	Unit of measurement
Focal length	2Z	100	mm
Radiation wavelength	λ	1064	mm
Diameter of the field of view	d	8	mm
Diameter of the focused beam	d_l	0,2...2,5	mm
The initial diameter of the radiation beam (after the lens) at the exp(-2) level	Ø	20	mm
The divergence of the beam (after the lens F= 1000) at the level exp(-2) - 1/e²	θ	~12	deg
Energy in a pulse	W_i	1	MJ
Radiation power	P_{csr}	6-7	W
Duration of the radiation pulse	t_i	30-70	ns
Operating frequency	F	10-30	kHz
Intensity (power density)	I	3*10⁵	W/cm²
Laserresonator:	τ	50	%
	τ	<1	%
	L	40	cm

Further, in order to develop the technological process of thermal hardening of the deposited layer by laser radiation, it is necessary to determine the optimal sample processing modes in which the qualitative characteristics of the deposited layer are obtained. In order to provide high-quality indicators of the deposited layer by laser hardening, it is required to determine the power density in the laser spot zone for metal fusion and obtaining an optimal (non-critical) penetration depth, surfacing thickness and width of the deposited coating. The penetration depth depends on the focal length and density of the material, due to the fact that a high heating of the transition region occurs in the focusing zone of the laser beam.

In section 3, the microhardness of the heat-strengthened layer of samples obtained by laser direction is investigated. According to the microhardness determination modes used, we determine the thickness of the surfacing from the focus of the laser beam; the width of the laser hardening zone depending on the surfacing speed, focal length and laser radiation power (Figure 54-56). The hardness of the deposited layer is determined by a portable combined MET-UDA hardness tester in the laboratory «Non-destructive testing and evaluation of physical and mechanical properties properties of parts» of the Department «Transport and Mechanical Engineering» of the M. Kozybaev Moscow State University. Table 12 presents the experimental optimal modes and qualitative parameters of the deposited layer.

Table 12. Qualitative characteristics and indicators of laser surfacing

Laserradiationpower, kW	Surfacingspeed, mm/s	Focallength, mm	Depositionthickness, mm	The width of the deposited layer, mm	Hardness, HV
1	10	10	0,87	10	585
1	10	15	0,78	10,2	572
1	12	19	0,69	11	565
1,5	13	20	1,0	11,5	541
1,6	14	22	0,81	12	573
1,7	15	30	0,9	12,5	595
2	16	35	0,96	12,9	584
2,02	17	36	1,01	13	595
2,4	19	40	1,29	13,8	604
2,6	20	55	1,35	14,1	624
3	20	60	1,45	14,3	641
3	20	70	1,48	15	617
3	20	80	1,61	15,1	622
3	20	85	1,47	15,5	610
3,1	20	100	1,56	16	661

Next, it is necessary to determine the regression equations of these dependencies according to the established experimental data.

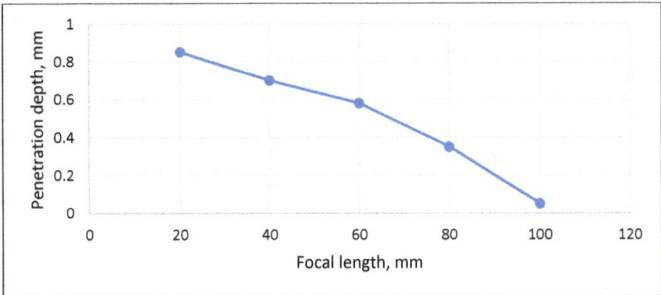

Figure 54. Change in the penetration depth depending on the amount of laser radiation focusing

According to the dependence of the penetration depth of the laser surfacing on the focusing distance, it can be established that the penetration depth of the laser radiation decreases with increasing focal length, and the following regression equation is obtained $y = -0,00001x^2 - 0,0009x + 0,2708$, $R^2 = 0,9983$.

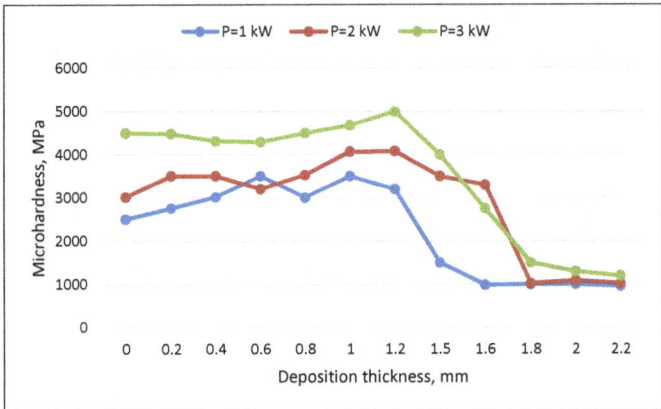

Figure 55. Dependence of the microhardness of laser radiation on the thickness of the deposited layer at different laser surfacing power

According to the obtained microhardness, depending on the thickness of the deposited layer at different laser deposition power, regression equations are established at 3 kW - $y = -60,327x^2 + 456,69x + 3839$, $R^2 = 0,8836$, at a power of 2000 watts - $y = -60,327x^2 +$

$456,69x + 3839$, $R^2 = 0,8836$, at 1000 watts - $y = -38,374x^2 + 274,14x + 2537,5$ with $R^2 = 0,7367$. It is established that a high microhardness of 5000 MPa is observed in a 0.8-1.5 mm thick section with a laser radiation power of 3 kW (Figure 57) with a regression equation $y = -0,45x + 1,9833$ and a correlation coefficient $R^2 = 0,9838$.

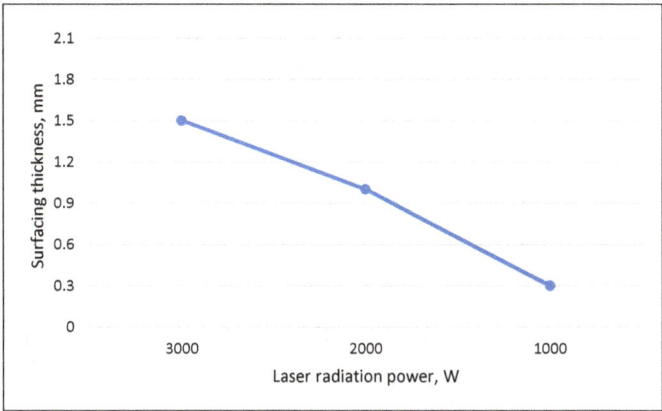

Figure 56. Change of laser surfacing power from the thickness of the laser radiation surfacing

The dependence of the change in the thickness of the laser direction, the location of the focal plane on the power of laser radiation at an optimal speed of 15 mm/s is shown in Figure 58. A decrease in the thickness of the deposited layer is shown with an increase in the focus of more than 80 mm of the laser spot. Regression equations are obtained at 3 kW - $y = -0,0357x^2 + 0,3083x + 1,064$ and $R^2 = 0,9576$, at 2 kW - $y = -0,0443x^2 + 0,3397x + 0,8$ and $R^2 = 0,9825$, at 1 kW - $y = -0,0493x^2 + 0,2987x + 0,65$ and $R^2 = 0,9993$.

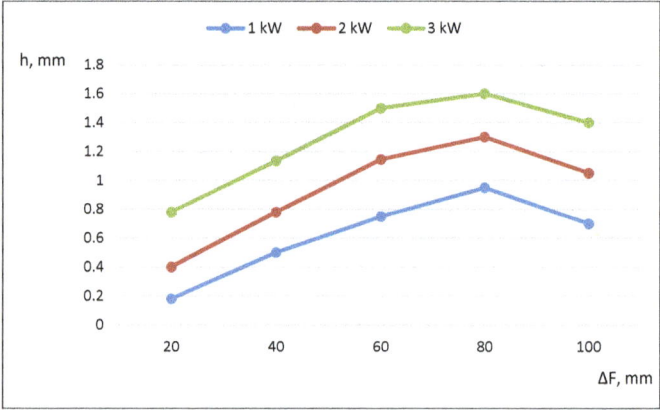

Figure 57. Dependence of the thickness of the laser deposition and the location of the focal plane on the power of laser radiation at an optimal speed of 15 mm/s

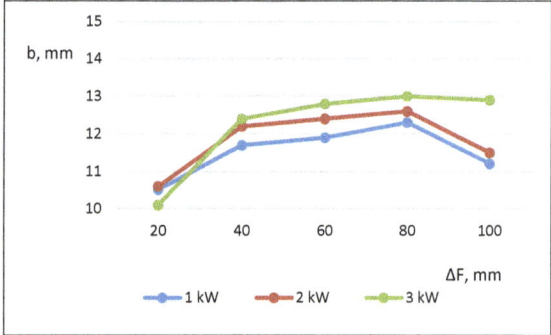

Figure 58. Dependence of the surfacing width and focal length on the laser surfacing power at a surfacing speed of 15 mm/s

According to this graph, we observe the optimal surfacing width of more than 10 mm to 13 mm at a power of 3000 W and defocusing - the optimal 40-80 mm. Regression equations of the dependence of the width of the laser exposure zone on the focal length at 3 kW are established - $y = -0{,}3571x^2 + 2{,}7629x + 7{,}88$ and $R^2 = 0{,}949$, at 2 kW - $y = -0{,}3857x^2 + 2{,}5343x + 8{,}5$ and $R^2 = 0{,}9607$, at 1 kW - $y = -0{,}3143x^2 + 2{,}0857x + 8{,}72$ and $R^2 = 0{,}9247$.

The average value of the hardness of the deposited layer is calculated by the equation (49):

$$HV = 1/n \sum_{i=0}^{n} X_i. \tag{49}$$

Therefore, when measuring the thickness and width of the laser exposure zone, the reproducibility dispersion is equal to $s_H^2 = 0{,}0081$ and $s_b^2 = 0{,}256$.

Analysis of the graphs shows that the optimal depth h of penetration is 0.75-0.8 mm, the maximum depth without crater formation is 0.95 mm at a power density of $3 \cdot 10^5$ W/cm^2. The thickness of the deposited layer is in the range of 0.1-0.3 mm in one pass with a focal length of 40-80 mm, an output power of up to 3 kW and a laser surfacing speed of 15 mm / s, achieves a high-quality hardened coating and technological modes of laser surfacing. Also, the quality of the deposited layer proves the correctness of choosing a solid-state pulsed laser. The total thickness of the deposited coating when surfaced with a pulsed laser should be in the range of 1.5 mm, which was obtained by us in 4-5 passes during the experiment and mathematical modeling. The next indicator of high-quality surfacing is the high adhesion of the wheel substrate and the filler material, as well as the absence of porosity in the fusion bath. The selected additive material in the form of a self-fluxing powder with the addition of alloying elements made it possible to obtain a hardened coating with better physico-chemical properties than the metal of the wheel substrate. Consequently, an increase in wear resistance of about 2 times, an increase in the hardness of the hardened coating up to 40 HRC compared to the initial hardness of 25-30 HRC (an increase to ≈5 HRC) has been established.

Further, the research is aimed at developing a technological process of energy-efficient technology for restoring wagon wheels. The final technological process (Figure 59) has predominant differences, such as the restoration of wagon wheels by laser surfacing, which is carried out on the developed mobile complex with the restoration of the original wheel geometry and the proposed optimal laser surfacing modes provide the restored wheels with wear resistance, erosion resistance, strength, elasticity and hardness close in value to the nominal values of the manufacturer. The described technology makes it possible to obtain the necessary operational durability, reliability and efficiency of restored wheel sets of railway production.

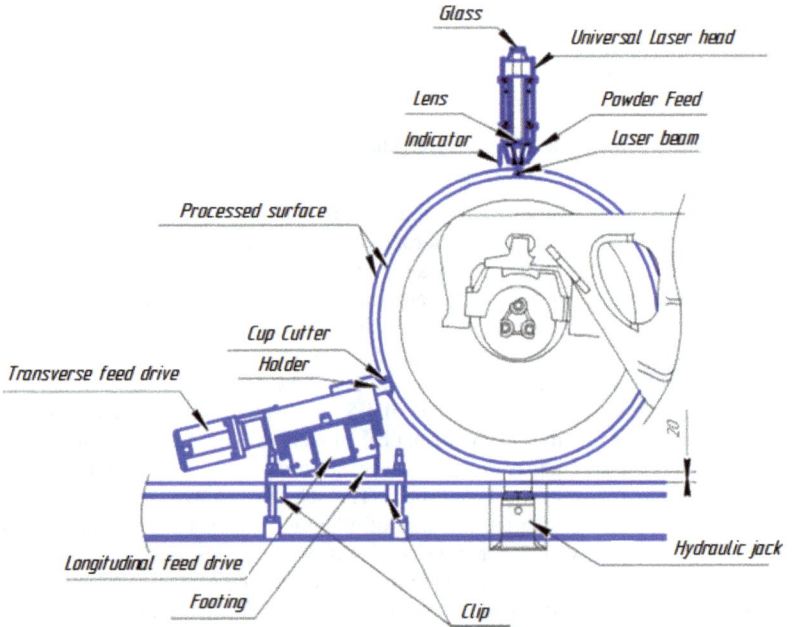

Figure 59. Technology of processing and restoration of a railway wheel

The technological process of restoring the worn surface of the wheel by laser deposition is as follows. Before the hardening process by laser surfacing, it is first necessary to grind the wheel to a complete sample of the defective area, then weld it taking into account the allowance for the final machining and finally grind it to the nominal size.

For longitudinal and transverse turning, when grinding an unacceptable defect, such as a notch (GOST 33788-2016), with a depth of 10 mm, the cutting speed is calculated by equality

$$V = \frac{C_v}{T^{mv} \cdot t^{xv} \cdot S^{yv}} \cdot k_v, \tag{50}$$

where $C_v; mv; xv; yv$ - empirical coefficient and degree indicators;

  T - the period of durability of the cutting tool;

  t - cutting depth, mm (equal to 3 mm – before surfacing; up to 1 mm – after surfacing);

  S - feed, mm/rev (1,1 – before surfacing; 1,5 – after surfacing);

k_v – correction factor.

$$k_v = K_{Mv} \cdot K_{Sv} \cdot K_{Tv}, \tag{51}$$

where K_{Mv} - a coefficient that takes into account the influence of the processed material

$$K_{Mv} = k_G \left(\frac{750}{\sigma_v}\right)^{nv}, \tag{52}$$

where K_{Sv} – surface condition coefficient;

K_{Tv} – the coefficient of action of the tool material;

σ_v - the actual parameter characterizing the processed material for the grade 2 wheel steel is equal to σ_v=910-1110;

k_g - coefficient characterizing a group of steel by machinability;

nv - degree indicator during processing.

Therefore, when turning preliminary and final, the turning parameters are equal (Table 13):

- cutting speed before laser surfacing (cutting depth – 3 mm; feed – 0.8-1.1 mm/rev) – 77 m/min;

- cutting speed after laser surfacing (cutting depth – from 0.3 to 0.5 mm; feed – 0.1 mm/rpm) – 67 m/min;

- rotation speed before laser surfacing – 25 rpm, after surfacing – 22 rpm.

Table 13. Parameters of wagon wheel turning

Type of turning	Feeds, mm/tur	Rotation speed n, tur/min	Cuttingdepth$f$,mm
Preliminary (beforelasersurfacing)	0,8-1,1	25	3
Final (finishing, after laser surfacing)	0,1	22	0,3-0,5

The final profile after laser deposition is shown in Figure 60, the optimal angle of inclination of the ridge, the width of the ridge and the chamfer angle are determined. At the same time, regardless of the amount of wear on the rolling surface of the opposite wheel, the same amount of metal must be removed on the opposite wheel (the diameters of the opposite ears must be the same).

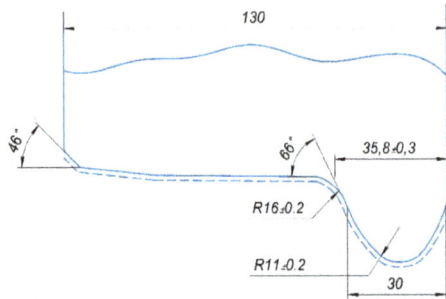

Figure 60. Wheel profile after laser deposition

Thus, the developed laser technology is energy efficient, increasing the durability and wear resistance of wagon wheels, the additive material used in the form of PG-CP2 powder gives a hardness of about HRC 40, also selected surfacing modes give and retain the shapes and sizes of the deposited layer and its physical and mechanical properties. The optimal modes of laser surfacing of a wagon wheel are substantiated: power density I, surfacing speed v_h and pulse duration t.

Conclusion

An important difference between laser surfacing and traditional surfacing methods is the precise and purposeful application of the filler material to the welded wheel with limited heat input, the hardened layer has a smaller microstructure with a small porosity, increasing corrosion resistance. The hardness of the deposited coating increases $HRC \geq 2 \div 4$, which is unattainable with classical methods of thermal hardening.

The dependences of the influence of the thickness h and width b of the hardened layer zone on the laser radiation power and the location of the focal length at the deposition rate of 10-20 mm/s are established. The dependences of the quality indicators of the deposited layer on the parameters of the technological regime are established, regression equations of the dependences of the width of the laser exposure zone (b), the thickness of the deposited layer (h) and the hardness of the hardened layer (HV) on the laser radiation power (P), the speed of the surfacing process (v) and the focal length (ΔF) are determined.

The established basic parameters of laser surfacing radiation power P and spot diameter d_p determine the power density, the duration of laser exposure depends on the speed of movement of the wheel relative to the laser beam. Laser deposition provides multiple improvements in the mechanical and tribological characteristics of the treated surface of the wheel.

The research results presented in the monograph were prepared during the implementation of the grant project of the Ministry of Science and Higher Education on the topic: IRN AR14869177 "Development and implementation of a new high-performance mobile railway complex with resource-saving technology for laser restoration of railway wheel sets".

References

[1]   GOST 10791-2011. The wheels are solid-rolled. Technical conditions. – With correction. from 29.12.2016; introduction. 01.01.2012. – Moscow: Publishing House of Standards, 2011. – 28 p.https://files.stroyinf.ru/Data2/1/4293800/4293800552.pdf

[2]   Kaspakbayev K.S., Estemirova R.S. Organization of the transportation process of railway transport of the Republic of Kazakhstan // Materiłyxv międzynarodowej naukowipraktycznej konferencji nauka i inowacja. Przemyśl Polska, 2019. - Vol. Pp. 70-76

[3]   Shibeko R. V., Zakharov E. A. Control system of wheel sets of railway wagons // Young scientist. - 2014. - № 18 (77). - pp. 314-317. https://moluch.ru/archive/77/13293/

[4]   Leonenko E. G. Interaction of the track and empty freight cars when moving in straight and curved sections of the track // Modern technologies. System analysis. Modeling. – 2019. – Vol. 63. - No. 3. – pp. 148-154.

[5]   Shevchenko D. V. Development of new methods for determining the power factors of the impact of rolling stock on the track / D. V. Shevchenko, R. A. Savushkin, Ya. O. Kuzminsky, T. S. Kuklin, E. A. Rudakova, A.M. Orlova // Railway equipment. – 2018. – № 1 (41). – Pp. 38-51.

[6]   Gubenko S. I. The influence of steel quality on the fatigue strength of solid-rolled wheels / S. I. Gubenko, I. A. Ivanov, D. P. Kononov // Factory Laboratory. Diagnostics of materials. – M.: Publishing house «Test-zl», 2018. – Vol. 84. – No. 3. - Pp. 52-60. https://doi.org/10.26896/1028-6861-2018-84-3-52-60

[7]   Ensuring the safe operation of wheels // Railways of the world. 2018. - No. 1. - pp. 59-63.

[8]   Maksimov I. N. Profile of the rolling surface of wheels for high-speed trains//Rail transport. 2014. No. 11. pp. 50-52.

[9]   Zhumekenova Z.Zh., Seitova A.T. Machine zhasaudagy standarttau zhane sapany baskaru.Petropavlovsk: IPO of M. Kozybayev NCSU, 2021. p. 101.

[10]   Khabirova S. Problems of repair of rolling stock // RZD-Partner. – 2006. – pp. 114-120.

[11]   Savoskin A. N., Vasiliev A. P. Dislocation model of wheel-rail interaction in the implementation of torque and lateral vibrations of crews // Izvestia PGUPS. 2017. - No. 1. - Pp. 103-109.https://doi.org/10.26518/2071-7296-2021-18-2-168-179

[12]   Kohanovskiy V. A., Glazunov D. V. Control of the characteristics of lubricants // Russian engineering research. 2017. - Volume 37. - No. 9. Pp. 768-773. https://doi.org/10.3103/S1068798X17090131

[13]   A.S. II84699, MKI V 60 V 9/12. Elastic wheel for rail carriage/A.L. Golubenko, A.S. Filonov, N.N., Kalyuzhny, A.N. Konyaev, V.P. Tkachenko, N.M. Kramar, I.N. Sukhov; Voroshilovgr. mashinostr. in-t and Voroshilovgr. teplovozostr. z-d. – Priority 15.06.

[14]   Lin F., Don H., Wang Y. Multipurpose optimization of the CRH3 EMU wheel profile // Adv Mech Eng. 2015. - Volume 7, 1-8. https://doi.org/10.1155/2014/284043

[15]   Bludov, A. N. The device of the operative non-contact diastolic system for preventing colonization of the non-road glands / A.N. Bludov, M.S. Chepchurov, E.M. Zhukov // Mechanics of the XXI century: collection of scientific tr. proceedings - 2014. - No. 13, 139-144.

[16]   Kononov D. P. Improving the reliability of solid–rolled wheels / D. P. Kononov. - M.: Publishing house «BIBLIO-GLOBUS», 2018. - 250 p.https://elibrary.ru/item.asp?id=32377309

[17]   Paul Molyneux-Barry, Claire Davis and Adam Bevan. The influence of wheel-rail contact conditions on the microstructure and hardness of railway wheels // Journal Scientific World. - Volume 2014. - Article ID 209752, 16 p.https://doi.org/10.1155/2014/209752

[18]   Belolapotkov D.A. Improving the accuracy of active control of the dimensions of parts in the manufacturing process / D.A. Belolapotkov, I.R. Dobrovinsky, Yu.T. Medvedik // The World of Measurements. - 2007. - No. 7. - Pp. 43-46. – ISSN 1813-8667.

[19]   Shur E. A. To the question of the optimal ratio of the hardness of rails and wheels // Modern problems of interaction of rolling stock and track: materials of scientific and practical conference. VNIIZHT. M., 2003. Pp. 87-93

[20]   Instructions for the maintenance of wagons in operation. Approved by the Council for Railway Transport of the state party. the commonwealth. Protocol No. 50 of May 21-22, 2009.

[21]   Korn G. and Korn T. Handbook of Mathematics for researchers and Engineers. St. Petersburg, publishing house «Lan», 2003, 832 p.http://engjournal.ru/catalog/mech/mdsb/1818.html

[22]   Norms of calculation and design of railcars of the MPC gauge of 1520 mm (non-self-propelled) with amendments and additions [Electronic resource]. Moscow, Publishing House of the Research Institute of Railway Transport, 1996, 318 p.

[23]   GOST 33783-2016. Wheel pairs of railway rolling stock. Wheel pairs of railway rolling stock. Methods for determining strength indicators. Moscow, Publishing house Standartinform. - 2016. - 68 p.

[24]   Ustich P.A., Karpychev V.A., Ovechnikov M. N. Reliability of rail non-traction

rolling stock. Moscow, Publishing House of the Educational and Methodological Center of the Ministry of Railways of Russia. - 2004. - 416 p.

[25] GOST 10791-2011. Kolesa tsel'nokatannyye. Technical conditions. Moscow, Standartinform Publ. - 2012. - 53 p.

[26] Kononov D.P. Improving the reliability of solid-rolled wheels. Moscow: Publishing House Biblio-Globus, 2018. - 250 p.

[27] Kiselev I.P. High-speed rail transport and prospects for its development in the world // Transport of the Russian Federation. Journal of Science, Practice and Economics. - 2012. - № 3-4 (40-41). - pp. 61-65.

[28] The effect of lubrication on the interaction of rolling stock and track// Railways of the world. 2005. - No. 9. - pp. 74-78.

[29] Ermolov I.N., Lange Yu.V. Non-destructive testing: handbook: in 7 volumes /edited by V.V. Klyuev. Vol.3: Ultrasonic control - M.: Mechanical Engineering, 2004. - 864 p.

[30] Patent 2191376 Russian Federation, IPC G01N29/04 Method for measuring the size of defects during ultrasonic inspection of products/ Chapaev I.G., Zhukov Yu.A., Luzin A.M. et al.; publ. 10/20/2002.

[31] Patent 2191376 Russian Federation, IPC G01N 29 Method of measuring the size of defects during ultrasonic inspection of products 04 / Chapaev I.G., Zhukov Yu.A., Luzin A.M. et al.; publ. 20.10.2002

[32] Budadin O.N., Potapov A.I., Kolganov V.I. et al. Thermal non–destructive testing of products. - M.: Nauka, 2002. - 476 p.

[33] Lazarenko A.P. Automatic detection of defects in radiation images of welds // Technical diagnostics and non-destructive testing. - No. 3, 2008. - pp. 31-37.

[34] Patent for utility model 131492 Russian Federation, IPC G01N 29/00 Automated ultrasound control system / Vopilkin A.X, Romashkin S.V., Tikhonov D.S.; publ. 20.08.2013.https://i.moscow/patents/ru145578u1_20140920

[35] Patent 2184373 Russian Federation, IPC G01N 29/04 Method of non-destructive testing of products / Markov A.A., Bershadskaya T.N., Belousov N.A.; publ. 27.06.2002

[36] Patent 139681 for utility model Russian Federation, IPC G01N29/04 Installation for contactless ultrasonic, and/or eddy current, and/or magnetic control of cylindrical products/Kirikov A.V., Borisov.N., V. Shcherbakov.A.; publ. 04/20/2014

[37] Klyuev, V. V. Globalization of technical diagnostics and non-destructive testing / V. V. Klyuev // Control. Diagnostics. - 2004. – No. 8. – pp. 3-6.

[38] Dudaeva L.G. Methods of non-destructive testing /Young scientist. – 2018. - № 34

(220). - pp. 6-10.

[39]   Patent 2539806 Russian Federation, IPC G01N29/04 Ultrasonic control for the
       protection of objects, ultrasonic transducer and ultrasonic council for the
       protection of objects / Inagaki Koichi, Izumi Mamoru, Karasavahirokazu; publ.
       27.01.2015

[40]   MorozovaT.Yu., Bekarevich A.A., Budadin O.N. A new approach to the
       identification of defects in materials// Control. Diagnostics, 2014. - No. 8. - pp. 42-
       48. https://doi.org/10.14489/td.2014.08.pp.042-048

[41]   Vaichunas G., Gelumbitskas G., Lingaitis L.P. Methods of studying the wear of
       locomotive axles. Transport problems: Transport problems. 2013. - Volume 8. -
       Issue 1. - pp. 95-103.

[42]   Harris W. J. Generalization of the best practices of heavy-weight movement:
       issues of wheel and rail interaction / W. J. Harris, S. Zakharov, J. Landgren, H.
       Tourne, V. Ebersen; translated from English; edited by S. M. Zakharov, V. M.
       Bogdanova. – M.: Intext, 2002. - 408 p.

[43]   Li Li. Optimal wheel profile design for a high-speed train / Li Li, K. Dabin, J.
       Xuesong. – Chengdu, China, 2015. - 22
       p.https://www.researchgate.net/publication/273498486_

[44]   Zakharov S. M. The development of the heavyweight movement in the world / S.
       M. Zakharov // VestnikVNIIZhT. – 2013. – No. 4. – pp. 9-
       17https://www.elibrary.ru/item.asp?id=19624056

[45]   Bogdanov, A.F. Operation and repair of wheel pairs of wagons / A.F. Bogdanov,
       V.G. Chursin - M.: Transport, 1985. – 270 p.

[46]   Mashnev, M.M. On the classification of defects of wheel pairs / M.M. Mashnev,
       R.S. Khrustalev // Railway transport. - 1968. - No. 2. - pp. 58-60. – ISSN 0044-
       4448.

[47]   Vohla, G.K.V. Werkzeugezurrealitaetsnahen simulation der laufdynamik von
       schienenfahrzeugen/ G.K.W. Vohla. – Fortschritt-Berichte VDI Reihe 1, Nr.270.
       VDI Verlag, Dusseldorf, 1998https://katalog.slub-dresden.de/id/0-12917047X

[48]   Research of solid-rolled wheels with stamped, rolled flat-conical and curved disc
       with the development of proposals: research report. - I560V-84. - Moscow:
       VNIIZhT, 79 p

[49]   Zhumekenova Z.Zh., Bondarev V.K. Types of defects in wheel sets of railway cars
       and ways to eliminate them. – Almaty: «Bulletin of KazNITU» No. 4, 2019. - pp.
       376-384

[50]   Shkarupa E.S. Wheel pairs: effective operation and repair/ E.S. Shkarupa // Wagon
       park. – 2007. - No. 2

[51]   Krutko Andrey A. Optimization of the technology of restorative turning of the

profile of the railway wheel / Andrey A. Krutko, Alexey A. Krutko // Young
Russia: advanced technologies – in industry. - 2013. - No. 1. - pp. 053-055.

[52]   Analysis of methods for restoring the profile of rolling wheel pairs / A.A.
       Vorobyev, I.A. Ivanov, D.P. Kononov // Bulletin of Scientific Research. Institute
       of Railway Transport. - 2011. - No. 3. - PP. 34-38. - ISSN 2223-9731

[53]   Kossov V. S. Investigation of the risk of freight cars derailment due to the
       destruction of one of the elements of the wheelset / V. S. Kossov, G. M.
       Volokhov, D. A. Knyazev // Labor safety in industry. - 2010. - No. 6. - pp. 42-46.

[54]   Becher S.A., Stepanova L.N., Kochetkov A.S. Development of a technique for
       rejecting defects of the rolling surface of wheel pairs in motion// Control.
       Diagnostics. - 2011. - No. 7. - pp. 24-29.

[55]   Kochetkov A.S., Becher A.S. Investigation of rail deformations for searching for
       surface defects of wagon wheels passed through it // Nauka. Industry. Defense:
       Proceedings of the VII All-Russian Scientific and Technical Conference -
       Novosibirsk: NSTU, 2006. - pp. 231-
       232.https://freereferats.ru/advanced_search_result.php?keywords=01005397205

[56]   Becher S.A., Kochetkov A.S., Kozyatnik I.I. Investigation of the distribution of
       deformations in the rail under a passing train to increase the reliability of detecting
       defects in the rolling surface of wagon wheels // Train safety: tr. VI scientific.-
       Practical conf. - M., 2005. - Vol. 2. - S. X-16 - X-17.

[57]   RomenYu.S. The condition of the running gear of rolling stock and wear in the
       wheel-rail system / Yu.S. Romen, A.M. Orlova, V.S. Forester // Bulletin of the
       Research Institute of Railway Transport, 2010. - No. 2. - pp. 42-
       45.https://www.dissercat.com/content/

[58]   PetrovS.Yu. The influence of lubrication on the wear of wheel ridges / S.Yu.
       Petrov, S.M. Baban, A.I. Kostyukevich // Locomotive, 2013. - No. 8. - pp. 43-45.

[59]   Orlova A.M. Clarification of some parameters of the wheel wear model of a
       freight car / Saidova A.V., Orlova A.M. // IzvestiyaPeterburgskogo University of
       Railways, 2013 - No. 1(34) – pp. 147-151.https://cyberleninka.ru/article/n/

[60]   Bogdanov V.M. Reducing the intensity of wear of wheel ridges and lateral wear of
       rails / V.M. Bogdanov // Railway Transport. - 1992. - No. 12. - pp. 30-34.

[61]   Bukharin M.N. We reduce lateral wear of rails and vertical undercutting of ridges /
       M.N. Bukharin // Locomotive. - 1993. - No. 8. - pp. 27-28

[62]   Bukhin M.V. Wagons with free-rotating wheels / M.V. Bukhin // Railway
       transport, 1965. - No. 2. - pp. 94-95.

[63]   Verigo M.F. Reasons for the increase in the intensity of lateral wear of rails and
       wheel ridges / Verigo M.F. - M.: Transport, 1992, 56 p.

[64]   Luzhnov Yu. M. Coupling of wheels with rails (nature and regularities) / Yu. M.

Luzhnov. – M.: In-text, 2003. – 144 p.

[65]   Shiler A.V. Investigation of dynamic properties of a wheelset with flexible independently rotating bandages / A.V. Shiler, V.V. Shiler, P. A. Shipilov // IzvestiyaTranssib. - 2011. - № 4 (8). - pp. 69-75https://cyberleninka.ru/article/n/

[66]   Shiler, A.V. Wheelset for railway rolling stock with flexible independently rotating bandages / A.V. Shiler, T.O. Bezugly // XII Tupolev readings: materials of the International Youth Scientific Conference – Kazan: KAI, 2004

[67]   Shiler A.V. Modeling and experimental study of the movement of a wheel pair with independent rotation of wheel circles / A.V. Shiler, V.V. Shiler, T.O. Bezugly, P.A. Shipilov, A.V. Ploskov // Resource-saving technologies on railway transport: collection of tr. of the All-Russian Scientific-Technical. conf. – Krasnoyarsk, 2005

[68]   Shiler A.V. Improving the energy efficiency of the production activities of JSC «Russian Railways» through the introduction of new technical and technological solutions/A.V. Shiler, E.S. Prokofieva, V.V. Shiler//Electronics and electrical equipment of transport. – 2018. – No. 6. – pp. 2-4

[69]   Pevsner V.O. Influence of gauge width/V.O. Pevsner //Railway transport. – 1996. – No. 12. – pp. 36-39

[70]   Harris W. J. Generalization of the best practices of heavy-weight movement: issues of wheel and rail interaction: Translated from English/ W. J. Harris, S. M. Zakharov, J. Landgren, H. Tourne, V. Ebersen. - M.: Intext, 2002. - 408 p.https://www.centrmag.ru/catalog/product/

[71]   Kragelskyand V. Frictional self-oscillations / And V. Kragelsky, N. V. Gittis. - M.: Nauka, 1987. - 171 p.

[72]   Buynosov A.P. The influence of wheelset diameter difference on their wear taking into account the means of technical diagnostics / A.P. Buynosov, K.A. Statsenko // Collection of works «Resource-saving technologies in railway transport», CHIPS UrGUPS. - Chelyabinsk, 2002. - pp. 24-33

[73]   Buynosov A.P. Wear of bandages and rails: causes and possibilities of reduction // Railway transport. - 1994. - No. 10. - pp. 39-41

[74]   Buynosov A.P. Wear of bandages and rails: causes and possibilities of reduction / A.P. Buynosov // Railway transport. - 1994. - No. 10. - pp. 39-41

[75]   Krysanova L.G. Improving the reliability of the upper structure of the track in modern operating conditions. - M.: Intext, 2000. - 142 p.http://geotm.dp.ua/index.php/ru/collection/54

[76]   Improving the interaction of rolling stock and track. // Railways of the world. 2004. - No. 8. - pp. 63-68

[77]   Heyman X. The direction of railway carriages of rail gauge / H. Heyman. M.:

Transzheldorizdat, 1957. - 415 p.

[78]   Optimization of wheel-rail interaction // Railways of the world. - 2003. - No. 1. - pp. 66-70

[79]   Zhumekenova Z., Savinkin V., Seitova A., Abilmazhinova A. Investigation of the reasons of resource longevity decrease of railway cars' wheel pairs. Scientific journal «Bulletin of KazNTU». –Almaty: No.2 (138) April 2020. - pp. 239-245

[80]   Kondrashov V.M. Unified principles of the study of the dynamics of railway carriages in theory and experiment // Scientific works of the All-Union Scientific Research Institute of Railway Transport. – M.: Intext, 2001. – 188 p.

[81]   Kossov V.S. Modeling of the energy interaction of a locomotive and a track for various tribological conditions of wheels and rails // Vestnik VNIIZhT. - M. - 2001. - No. 2. - pp. 17-19

[82]   Kossov V.S. Results of dynamic and on the impact on the path of tests of trains of increased mass and length / V.S. Kossov, V.A. Gapanovich, A.A. Lunin, A.V. Spirov, A.V. Trifonov // Technique of railways. - 2018. - No. 2. - pp. 82-87

[83]   Kossov V.S. The influence of width tracks on indicators of dynamics, impact on the track, resistance to movement and criteria for wear of rails / Kossov V.S., Bidulya A.L., Berezin V.V. Bykov V.A., Lunin A.A., Grinevich V.P., Spirov A.V., Trifonov A.V. // Vestnik VNIKTI. - Kolomna, 2010. - Issue 92. - pp. 3-21

[84]   Adams Mechanical Dynamics / Corporate User Manual, Ann Arbor. - Michigan, USA, 2002. - 64 p.https://help-be.hexagonmi.com/bundle/

[85]   Eadie D.T., Kalouser J. Sray it on, let'em roll // Railuay Agc. – 2001. – № 6. – pp. 48–49

[86]   Filippov, V. N. Reduction of undercutting of ridges of wheel sets of freight cars / V. N. Filippov, P. I. Zuikov, Ya. D. Podlesnikov // Rail transport. - 2014. - No. 3. - pp. 70-72

[87]   Contact tasks of railway transport / V. I. Sakalo, V. I. Kossov. – M.: Mashinostroenie, 2004. - pp. 448-449

[88]   Classification of works in the field of computational and experimental methods for determining the wear of the profiles of the wheels of freight cars / Orlova A.M., Saidova A.V. // Tez. reports of International Scientific-Technical. conf. «Rolling stock of the XXI century: ideas, requirements, projects», St. Petersburg, 06.07-10.07.2011. -St. Petersburg: PGUPS, 2011. - pp. 49-50

[89]   Elastic deformation and the laws of friction / J. E. Archard // Proc. Royal Society. - London, 1957. - Ser. A243. – pp. 190-205. https://doi.org/10.1098/rspa.1957.0214

[90]   Dumpala, R.; Chandran, M.; Rao, M. S. Engineered CVD Diamond Coatings for Machining and Tribological Applications. JOM. 2015, 67, 7, 1565–1577.

https://doi.org/10.1007/s11837-015-1428-2

[91] Kuznetsov V.M. About the pointed coasting on the ridges of wheel sets // Path and track economy. 2000. - No. 9. - pp. 16-19

[92] Development of mathematical models of wagons on trolleys 18-9810 and 18-9855 for the study of wheel wear / Saidova A.V., Orlova A.M. // Tez. reports of the XIII International conference "Problems of mechanics of railway transport. Traffic safety, dynamics, strength of rolling stock and energy saving", Dnepropetrovsk, 23.05-25.05.2012. - Dnepropetrovsk: Dnepropetr. nats. Univ. zh.-d. transp. named after academician V. Lazaryan, 2012. - pp. 128-129

[93] Sakalo V. I. Contact problems of railway transport / V. I. Sakalo, V. S. Kossov. - M.: Mashinostroenie, 2004. - 496 p.

[94] Yershkov O. P. Issues of railway track preparation for high traffic speeds. - Moscow: Transzheldorizdat, 1959. - 126 p.

[95] Karpushchenko N.I., Kotova I.A. Wear and service life of rails and wheels of rolling stock. SibGUPS, Russia. UDC 625.1.03. - pp. 41-46. https://doi.org/10.15802/stp2003/21174

[96] Voronko, A. N. Analysis of criteria for stability of railway crews from derailment / A. N. Voronko, S. Yu. Sapronova, V. P. Tkachenko // Bulletin of VNU named after V. Dahl. - 2006. - № 8 (102), Part 1. - pp. 115-120

[97] Sakalo, V. I. Contact problems of railway transport / V. I. Sakalo, V. S. Kossov. - M.: Mashinostroenie, 2004

[98] Modeling of contact interactions in problems of dynamics of systems of bodies /D. Yu. Pogorelov et al. // Dynamics, strength and reliability of transport machines: Collection of scientific tr. / edited by V. I. Sakalo. - Bryansk: BSTU, 2001. - pp. 11-23

[99] Manashkin, L. A. Vibration dampers and shock absorbers of rail carriages (mathematical models): monograph / L. A. Manashkin, S. V. Myamlin, V. I. Prikhodko. - D.: ART-PRESS, 2007. - 196 p. https://doi.org/10.15802/978-966-348-121-0

[100] Degtyareva L.N., OseninYu.I., Myamlin S.V. Mathematical description of the force interaction of wheels and rails. UDC 629.4.067. - pp. 21-24

[101] Myamlin, S. V. Modeling the dynamics of rail carriages. - D.: New Ideology, 2002. - 240 p.

[102] Zhumekenova Z.Zh., Savinkin V.V., Kolisnichenko S.N. On the issue of promising technologies for restoring wear surfaces. "Bulletin of KazNITU". – Almaty: No.2 (138), April 2020. - pp. 170-177

[103] Vorobyev A.A., Ivanov I.A., Kononov D.P., etc. Analysis of methods for restoring the profile of rolling wheel pairs // Bulletin of the Research Institute of Railway

Transport. - 2011. - No. 3. - pp. 34-38

[104] Buynosov A.P.//Bulletin of Transport of the Volga region. 2010. - № 4(24). - pp. 21-25

[105] Buynosov A.P., Pyshny I.M. // Scientific and Technical Bulletin of the Volga region. 2012. - № 2. - pp. 122-126

[106] Krutko A.A. Optimization of the technology of restorative turning of the profile of the railway wheel // Young Russia: advanced technologies – in industry. - 2013. - No. 1.

[107] Zhumekenova Z.Zh., Abilmazhinova A.S., Seitova A.T. Modern technologies of restoration of wagon wheels. Proceedings of the VIIIth International Scientific and Practical Conference «Science and Education in the modern world: challenges of the XXI century». - - Nursultan, 2020. - pp. 233-237

[108] Patent 2424091 Russian Federation, IPC B 23 To 9/04, B 23 To 35/36, C 22 B 9/18. Flux for electroslag welding or surfacing during the restoration of parts or electroslag remelting / E. G. Babenko, E. N. Kuzmichev, E. A. Drozdov, M. A. Kolesnikov. – No. 2009125939/02; publ. 20.07.2011. byul. No. 20

[109] Wieczorek, A.N.; Stachowiak, A.; Zwierzycki, W. Prediction of tribocorrosive properties of ADI containing Ni-Cu-Mo. Archives of Metallurgy and Materials. 2018, 63, 3. - pp. 1417-1422

[110] Wieczorek, A.N.; Stachowiak, A.; Zwierzycki, W. Experimental determination of the synergistic components of tribocorrosive wear of Ni-Cu-Mo-Ausferritic Ductile Iron. Archives of Metallurgy and Materials. 2018, 63, 1. - pp. 87-97.

[111] Gerasimova, A. A.; Keropyan, A. M.; Girya, A. M. Research of the Wheel-Rail System of Quarry Locomotives during the Traction Mode. Problems of Mechanical Engineering and Machine Reliability. - 2018, 1. -P. 39-42. https://doi.org/10.3103/S1052618818010065

[112] Savinkin V., Zhumekenova Z., Sandu A., Vizureanu P., Savinkin S., Kolisnichenko S., Ivanova O. Study of wear and redistribution dynamic forces of wheel pairs restored by a wear-resistant coating 15Cr17Ni12V3F. - Coatings 2021, 11(12), https://doi.org/10.3390/ coatings11121441

[113] Hesam, S.; Majid, M. Tribological Aspects of Wheel-Rail Contact: A Review of Wear Mechanisms and Effective Factors on Rolling Contact Fatigue. UrbanRailTransit. - 2017, 3. - pp. 227-237. https://doi.org/10.1007/s40864-017-0072-2

[114] Zhumekenova Z.Zh. Analysis of the most frequent defects of wheel sets of railway cars. Materials of the VI International Student Scientific and Practical Conference «Youth and Science - 2019», M. Kozybayev NCSU. - Petropavlovsk, 2019

[115] Zhumekenova Z.Zh., Abilmazhinova A.S., Seitova A.T. Zhylzhymaly

kuramgakyzmet korsetu zhane zhondeu zhuyesindegi resource unemdeutechnologiyalary. Materials of the international scientific and practical conference «Kozybayev readings - 2020: priority directions of development, achievements and innovations of modern Kazakh science». – Petropavlovsk, 2020, III. - pp. 275-279.

[116] Savinkin V.V., Kolisnichenko S.N., Kolisnichenko S.V., Zhumekenova Z.Zh. Investigation of the dynamic model of the crank-slide mechanism of piston pumps of the drilling complex. Materials of the international scientific and practical online conference «Youth and Science - 2021», No. IV. –Petropavlovsk. - pp. 377-379

[117] Savinkin V.V., Kolisnichenko S.N., Sandu A.V., Ivanova O.V., Petrica Vizureanu, Zhumekenova Z. Zh. Investigation of the strength parameters of drilling pumps during the formation of contact stresses in gears. Applied Sciences (Switzerland), 2021, 11(15), 7076. https: //doi.org/10.3390/app11157076.

[118] Patent 5935 RK. Mobile repair complex for the restoration of wheel sets of railway cars / Savinkin V.V., Shagaev I.V., Zhumekenova Z.Zh.; publ. 19.03.2021, Bul. No. 11. - 6 p.

About Authors

Zaure Zhetpisbaevna Zhumekenova

zzhzhumekenova@ku.edu.kz

Associate Professor of the Faculty of Engineering and Digital Technologies of the Department of Transport and Mechanical Engineering of M. Kozybayev NKU, Kazakhstan

Researcher at the National Center for State Scientific and Technical Expertise

PhD Zaure Zhetpisbaevna Zhumekenova is an associate professor of the Faculty of Engineering and Digital Technologies of the Department of Transport and Mechanical Engineering of the M. Kozybaev NKU in Petropavlovsk. In 2022, she defended her doctoral dissertation in mechanical engineering in the field of restoration of wagon wheels of rolling stock. She has published more than 20 scientific articles, some of which are indexed by SCOPUS and ISI Web of Science. She is the author of 1 patent on the topic of dissertation. She is a researcher at the National Center for State Scientific and Technical Expertise in the field of mechanical engineering, car building. Her main field of activity is mechanical engineering with involvement in solving problems of equipment wear, outdated technologies for the restoration of parts.

Vitaliy Vladimirovich SAVINKIN

https://ku.edu.kz/   cavinkin7@mail.ru

Professor at *Faculty of Engineering and Digital Technologies,* Department of «Transport and Mechanical Engineering» of the M. Kozybayev North Kazakhstan University, Kazakhstan

Chief Researcher of the project of the Ministry of Science and Higher Education of the Republic of Kazakhstan, associate Professor

Doctor of Technical Sciences Vitaly Vladimirovich SAVINKIN - Professor of the Faculty of Engineering and Digital Technologies of the North Kazakhstan University named after M. Kozybayev. Since 2016, he has defended his doctoral dissertation on road construction and lifting machines with the invention. In the same year 2016, by order of the Ministry of Education and Science of the Republic of Kazakhstan, he was approved as a member of the editorial board of the scientific and technical journal "Metrology". Since 2017, he has been a member of the editorial board of the European Journal of Materials Science and Engineering (EJMSE) published by The "Gheorghe Asachi" Technical University of Iasi (Romania) indexed in the international database SCOPUS and Web of Science. According to the results of the research, more than 120 scientific papers have been published, including 6 monographs, 3 of them in the editorial office of the USA, more than 12 articles in foreign publications indexed in SCOPUS, 40 articles in publications recommended by the KKSON of the Republic of Kazakhstan and the Higher Attestation Commission of the Russian Federation, 11 patents for utility models and inventions of the Republic of Kazakhstan and the Russian Federation, 30 works in materials of international scientific conferences and other publications. Savinkin V.V. has published 12 scientific articles in international journals indexed in the international database SCOPUS Q2 (Percentile according to CiteScore Scopus-71), has a Hirsch index h=5.

His main field of activity is innovative technologies for modifying the physical and mechanical properties of the material during the restoration of energy and mechanical engineering parts.

Andrei Victor SANDU

Associate Professor at Faculty of Materials Science and Engineering, "Gheorghe Asachi" Technical University of Iasi, Romania

Senior Researcher at National Institute for Research and Development for Environmental Protection INCDPM

Associate member of Romanian Academy of Scientists

President of Romanian Inventors Forum

http://afir.org.ro/sav/   andrei-victor.sandu@academic.tuiasi.ro

Dr.Eng. Andrei Victor SANDU is associate professor at Faculty of Materials Science and Engineering, Technical University "Gheorghe Asachi" of Iaşi. He has his PhD in Materials Engineering since 2012 with summa cum laudae. He has published over 450 scientific articles, over 400 indexed by SCOPUS and more than 300 indexed by ISI Web of Science. H-index is 28. He is co-author of 40 patents and other 10 patent applications (Romania, R. Moldova and Malaysia) and he has published 11 books, 4 of them in USA. He is Publishing editor for International Journal of Conservation Science (Web of Science and Scopus indexed) and European Journal of Materials Science and Engineering, also a reviewer for more than 20 Web of Science indexed journals. He is visiting professor at Universiti Malaysia Perlis and also President of Romanian Inventors Forum. Based on his expertise he is also Senior Researcher for National Institute for Research and Development for Environmental Protection INCDPM and Representative for Romania at IFIA (International Federation of Inventors' Associations) and WIIPA (World Invention Intellectual Property Associations. His main field is materials science with involvement in environmental issues, advanced characterization and obtaining of geopolymers and biomaterials.

Petrică VIZUREANU

www.afir.org.ro/peviz   petrica.vizureanu@academic.tuiasi.ro

Professor Eng. Ph.D., M.Sc. at "Gheorghe Asachi" Technical University of Iasi,

Faculty of Materials Science & Engineering

Director of the Department of Technologies and Equipment for Materials Processing, Faculty of Materials Science & Engineering

Vizureanu Petrica (ORCID: 0000-0002-3593-9400), is Full Professor on "Gheorghe Asachi" Technical University of Iasi with a teaching experience over 30 years, a very rich experience in project management of a national and international research projects (Director – 7, member – 35), concerns concretized in many articles in different competences areas (over 240 articles with >1500 citations Web of Science, H-index = 24): analytical methodologies for application in environmental chemistry field, elaboration of innovative technologies for water filtration/decontamination, and waters quality assessment, geopolymers, metallic biomaterials, computer assisted design, safety and health at work, management and commercial engineering, materials science. Author of 28 national books and 19 international chapters/books. Editor of books INTECH OPEN (10 books + 9 chapters).

www.ingramcontent.com/pod-product-compliance
Lightning Source LLC
Chambersburg PA
CBHW071717210326
41597CB00017B/2516